MODERN PHYSICS:

AN INTRODUCTORY SURVEY

This book is in the
ADDISON-WESLEY SERIES IN PHYSICS

ARTHUR BEISER

MODERN PHYSICS:

AN INTRODUCTORY SURVEY

ADDISON-WESLEY PUBLISHING COMPANY
READING, MASSACHUSETTS
MENLO PARK, CALIFORNIA · LONDON · AMSTERDAM · DON MILLS, ONTARIO · SYDNEY

ISBN 0-201-00515-8
FGHIJKLMNO-CO-798765

Preface

As its title suggests, this book is a brief introduction to the fundamental ideas of modern physics. It is designed to supplement a conventional general physics text, so that a balanced course can be presented regardless of the emphasis of the primary text. However, it can equally well be used as the basis of an abbreviated modern physics course or for independent study. In an effort to make the book as self-contained as possible, a number of topics from mechanics, kinetic theory, and the electromagnetic theory of light have been included which, though not part of modern physics, nevertheless are relevant to it and are not always discussed to the required extent in general physics texts.

In *Modern Physics: An Introductory Survey* the concepts of modern physics are considered chiefly in relation to the insights they provide on the structure and behavior of matter. There are sections on molecules and solids as well as atoms and nuclei. Consistent with the level of the treatment, derivations and numerical examples are given wherever possible. Because the prospective audience is not confined to students majoring in physics and the only prerequisites are elementary physics and a smattering of calculus, it has not been possible to develop the techniques of quantum theory and quantum statistical mechanics in any detail, and some results accordingly have to be accepted on faith by the reader. Despite this, quantum notions are employed throughout along with indications of their significance and range of application, since these notions are central to nearly all of modern physics.

Ile des Embiez, France   A. B.
May, 1967

Contents

1

Electromagnetic Waves

"The most obvious mechanical phenomenon in electrical and magnetical experiments is the mutual action by which bodies in certain states set each other in motion while still at a sensible distance from each other. The first step, therefore, in reducing these phenomena into scientific form, is to ascertain the magnitude and direction of the force acting between the bodies, and when it is found that this force depends in a certain way upon the relative position of the bodies and on their electric or magnetic condition, it seems at first sight natural to explain the facts by assuming the existence of something either at rest or in motion in each body, constituting its electric or magnetic state, and capable of acting at a distance according to mathematical laws ... The mechanical difficulties, however, which are involved in the assumption of particles acting at a distance ... are such as to prevent me from considering this theory as an ultimate one, though it may have been, and may yet be useful in leading to the coordination of phenomena. I have therefore preferred to seek an explanation of the fact in another direction, by supposing them to be produced by actions which go on in the surrounding medium as well as in the excited bodies, and endeavouring to explain the action between distant bodies without assuming the existence of forces capable of acting directly at sensible distances. The theory I propose may therefore be called a theory of the *Electromagnetic Field*, because it has to do with the space in the neighborhood of the electric or magnetic bodies ... The general equations are next applied to the case of a magnetic disturbance propagated through a nonconducting field, and it is shewn that the only disturbances which can be so propagated are those which are transverse to the direction of propagation, and that the velocity of propagation ... is so nearly that of light, that it seems we have strong reason to conclude that light itself (including radiant heat, and other radiations if any) is an electromagnetic disturbance in the form of waves propagated through the electromagnetic field according to electromagnetic laws."

James Clerk Maxwell (1831–1879)

Relativity and quantum theory, the twin pillars of modern physics, both had their genesis in the inability of classical electromagnetic theory to account for certain crucial experiments involving electromagnetic waves. This failure was ironic, because electromagnetic theory was in many ways the most successful product of classical physics and James Clerk Maxwell's masterly prediction of electromagnetic waves and his realization of the nature of light provided the theory with seemingly invincible support. It is therefore appropriate to begin our study of modern physics with a recapitulation of some of the basic notions of electromagnetic waves and a look into one of its major failures, the aptly-named "ultraviolet catastrophe."

1–1 MAXWELL'S HYPOTHESIS

An electric field is invariably associated with a changing magnetic field. This effect is known as electromagnetic induction, and the consequent electric field is present regardless of whether the magnetic field variations occur in free space or in a material medium. There is also a converse process, in which a changing electric field produces a magnetic field—even in empty space, where electric currents cannot flow. No simple experiment can directly demonstrate the latter effect, and it was first proposed by Maxwell on the basis of somewhat indirect reasoning.

When Maxwell expressed mathematically what was known in his time about electricity and magnetism in terms of Faraday's intuitive ideas of fields of force, he found that the resulting equations became particularly simple and symmetrical if he assumed that a magnetic field arises from a varying electric field as well as from an actual electric current. This assumption is actually less arbitrary than it may seem. Suppose that we attach two metal plates to the opposite terminals of a battery, as in Fig. 1–1. At first currents flow in the wires, and electric charge of opposite signs accumulates on the plates. Ultimately enough charge is present for the potential difference between the plates to equal that between the terminals of the battery, and the currents cease. While charge builds up on the plates, the electric field around them changes accordingly: the larger the current, the greater the rate of change of the field, and the smaller the current, the less the rate of change of the field. The behavior of the current is thus mirrored in the *rate of change* of the electric field surrounding the termination points of the current. By asserting that a varying electric field produces a magnetic field, then, Maxwell was no more than extending the correspondence between electric currents and varying electric fields to include magnetic effects. That such an extension also simplified the equations of electricity and magnetism encouraged Maxwell to take this notion seriously, because fundamental physical principles usually (though not always) may be expressed in simple form.

It is not surprising that electromagnetic induction became known many years before its converse was suspected. Even a slight electric field causes a current to flow in a conductor, and, if the resistance of the conductor is low enough, the current may be sufficiently large to detect despite the feebleness of the field itself. However, electric current has no magnetic counterpart because single magnetic poles do not exist;

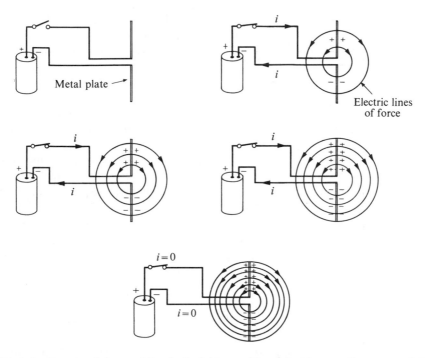

FIG. 1-1. The rate of change of the electric field around the plates of a capacitor corresponds to the flow of current: the larger the current, the faster the accumulation of charge, and consequently the greater the rate of change in the field.

the opposite poles of a magnet cannot be separated from one another the way opposite electric charges can. Weak magnetic fields are therefore hard to measure no matter where they occur, unlike weak electric fields, and those due to changing electric fields are seldom strong. Maxwell's notion that an electric field that varies with time gives rise to a magnetic field accordingly did not originate in an observation, but instead developed from an intuitive feeling for order in the natural world.

One of the tenets of the scientific method of inquiry is that all hypotheses must be capable of being experimentally verified, directly or indirectly, if they are to mean anything at all. Maxwell's next step was to seek a phenomenon that was a unique consequence of his hypothesis.

1-2 ELECTROMAGNETIC WAVES

It follows from electromagnetic induction that whenever there is a change in a magnetic field, an electric field is produced, and it follows from Maxwell's hypothesis that whenever there is a change in an electric field, a magnetic field is produced. Evidently it is impossible to have either effect occur alone. The electric field that arises from a change in a magnetic field is in itself a change in the pre-existing electric field (which might have had any original value, including zero), and therefore causes another magnetic field. The latter magnetic field, too, represents a change, and from the change an electric field is in turn produced. The process continues indefinitely, with a definite coupling between the fluctuating electric and magnetic fields. On the basis of his hypothesis, together with the other principles of electricity and magnetism, Maxwell was able to develop a detailed picture of how these field fluctuations travel through space.

The first idea that emerged from Maxwell's analysis was that the field fluctuations spread out in space from an initial disturbance in the same manner that waves spread out from a disturbance in a body of water; hence the name *electromagnetic waves* to describe them. If we throw a stone into a pond or otherwise alter the state of the water surface at some point, oscillations occur in which energy is continually interchanged between the kinetic energy of moving water and the potential energy of water higher than its normal level. These oscillations begin where the stone lands, and spread out as waves across the surface of the pond. The wave speed depends upon the properties of the pond water, varying with temperature, impurity content, and so on, but it is independent of the wave amplitude. This is typical wave behavior.

When electromagnetic waves spread out from an electric or magnetic disturbance, their energy is constantly being interchanged between the fluctuating electric field and the fluctuating magnetic field of the waves. Let us suppose that we have a periodic source of electromagnetic waves, say a pair of electric charges connected by a vibrating spring as in Fig. 1–2. In part (a) of the figure the charges are moving apart; the electric lines of force surrounding the charges are as shown, and the magnetic lines of force produced by the motion of the charges, which are concentric circles perpendicular to the paper, are indicated by crosses when their direction is into the paper and by circles when their direction is out of the paper. In (b) the charges have reached the limit of their motion and have stopped, so that they cease to produce a magnetic field. The outer magnetic lines of force do not disappear instantly because of the finite speed at which changes in electric and magnetic fields travel. If the charges remain at rest permanently, the entire magnetic field would in time vanish. However, as in (c), the charges now move toward each other, generating a magnetic field whose lines of force have the opposite direction to those created in (a). The electric lines of force have the same direction they had initially. In (d) the charges have passed each other and now are moving apart; the electric lines of force near them accordingly reverse their direction, but the magnetic lines of force are in the same direction as those produced in (c).

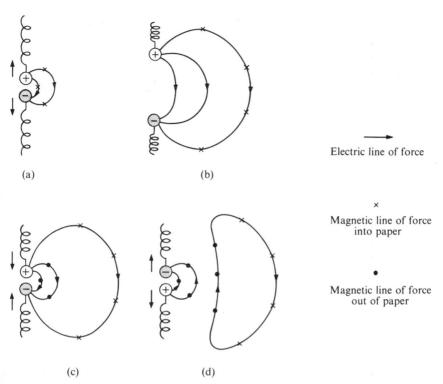

(a)   (b)

Electric line of force

×
Magnetic line of force
into paper

•
Magnetic line of force
out of paper

(c)   (d)

FIG. 1–2. A pair of electric charges that vibrate back and forth is a source of electromagnetic waves.

Owing to this sequence of changes in the fields, the outermost electric and magnetic lines of force respectively form into closed loops. These loops of force, which lie in perpendicular planes, are divorced from the oscillating charges that gave rise to them and continue moving outward, constituting an electromagnetic wave. As the charges continue oscillating back and forth, further associated loops of electric and magnetic lines of force are emitted, forming an expanding pattern of loops.

Figure 1–3 shows the configuration of the electric and magnetic fields that spread outward from a pair of oscillating charges. The actual fields are present in three dimensions, so that the magnetic lines of force form loops in planes perpendicular to the line joining the charges. There are three significant things we should note about these electromagnetic waves, which are as follows:

(1) The variations occur simultaneously in both fields (except close to the oscillating charges), so that the electric and magnetic fields have maxima and minima at the same times and in the same places.

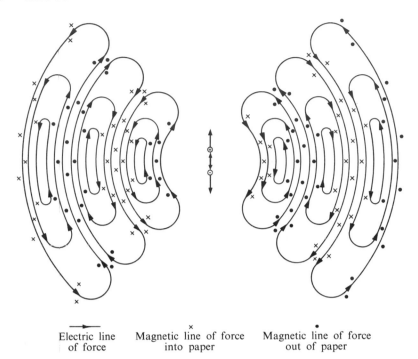

——→	×
Electric line of force	Magnetic line of force into paper

•
Magnetic line of force out of paper

FIG. 1-3. The configuration of the electric and magnetic fields that spread outward from a pair of oscillating charges.

(2) The directions of the electric and magnetic fields are perpendicular to each other and to the direction in which the waves are moving. Light waves are therefore transverse.

(3) The speed of the waves depends only upon the electric and magnetic properties of the medium they travel in, and not upon the amplitudes of the field variations.

Figure 1-4 is an attempt at portraying (1) and (2) of the above in terms of superimposed graphs showing the electric and magnetic field intensities and directions in an electromagnetic wave at a particular instant. The actual configuration of the fields in space is shown in Fig. 1-3. It is worth keeping in mind that, unlike the case of mechanical waves —water waves, sound waves, waves in a stretched string—nothing material moves in the path of an electromagnetic wave. The only changes are in field intensities.

Maxwell's theory of electromagnetic waves showed that their speed c in vacuum depends solely upon ϵ_0 and μ_0, the permittivity and permea-

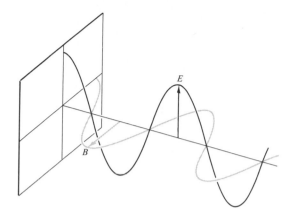

FIG. 1-4. The electric and magnetic fields in an electromagnetic wave vary simultaneously. The field directions are perpendicular to each other and to the direction of propagation.

bility of free space. Maxwell found that the speed c is given by

$$c = \frac{1}{\sqrt{\epsilon_0 \mu_0}} \qquad (1\text{-}1)$$

$$= \frac{1}{\sqrt{8.85 \times 10^{-12}\ \text{coul}^2/\text{n·m}^2 \times 1.26 \times 10^{-6}\ \text{w/amp·m}}}$$

$$= 3.00 \times 10^8\ \text{m/sec},$$

which is the same speed that had been experimentally measured for light waves in free space. The correspondence was too great to be accidental, and, as further evidence became known, the electromagnetic nature of light found universal acceptance.

While the correct value of c is

$$c = 2.998 \times 10^8\ \text{m/sec}$$

to four significant figures, we shall use the value

$$c = 3 \times 10^8\ \text{m/sec}$$

in what follows.

1-3 VARIETIES OF ELECTROMAGNETIC WAVES

Light is not the only example of an electromagnetic wave. While all electromagnetic waves share certain basic properties, other features of their behavior depend upon their frequencies. Light waves themselves span a brief frequency interval, from 3×10^{14} cycles/sec for red light to 8×10^{14}

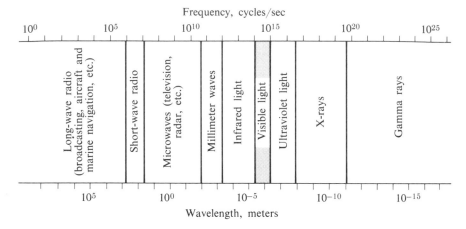

FIG. 1–5.   The electromagnetic wave spectrum.

cycles/sec for blue light. Electromagnetic waves with frequencies between these limits are the only ones that the eye responds to, and specialized instruments of various kinds are required to detect waves with higher and lower frequencies. Figure 1–5 shows the electromagnetic wave *spectrum* from the low frequencies used in radio communication to the high frequencies found in x-rays and gamma rays (about which we shall have more to say in later chapters). The wavelengths corresponding to the various frequencies are also shown.

As we recall, the product of the frequency f of a wave and its wavelength λ is just the wave speed, here c. Hence, given the wavelength or frequency of a particular electromagnetic wave, we can immediately find the complementary quantity. The wavelength of red light, for instance, is

$$\lambda = \frac{c}{f} = \frac{3 \times 10^8 \text{ m/sec}}{4 \times 10^{14} \text{ cycles/sec}}$$

$$= 7.5 \times 10^{-7} \text{ m},$$

which is less than 1/1000 of a millimeter. By contrast, a 1 megacycle/sec radio wave has a wavelength of 300 meters. (1 kilocycle = 10^3 cycles; 1 megacycle = 10^3 kilocycles = 10^6 cycles.)*

1–4 THE POYNTING VECTOR

Associated with every electric field of intensity E is an energy density $\frac{1}{2}\epsilon_0 E^2$ and with every magnetic field of flux density B is an energy density $\frac{1}{2}B^2/\mu_0$. Electromagnetic waves therefore contain energy, and

---

* The abbreviations for megacycle and kilocycle are mc and kc.

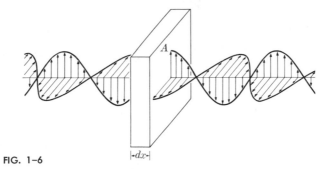

FIG. 1-6

their motion involves the flow of energy. The *Poynting vector* **S** describes this flow of energy. The direction of **S** is the direction of the electromagnetic wave in question, and its magnitude S is equal to the rate at which energy is being transported by the wave per unit cross-sectional area. We can derive a formula for S in terms of the simple situation of Fig. 1-6, which shows an electromagnetic wave passing through an imaginary box of area A and thickness dx. If E and B are the values of the electric field intensity and magnetic flux density of the wave in the box at any instant, the total energy density w in the box at that instant is

$$w = \frac{1}{2}\,\epsilon_0 E^2 + \frac{1}{2}\,\frac{B^2}{\mu_0}. \tag{1-2}$$

The total energy W in the box is

$$W = wV,$$

where V is the volume of the box. Since $V = A\,dx$, we have

$$W = \frac{1}{2}\left(\epsilon_0 E^2 + \frac{B^2}{\mu_0}\right) A\,dx. \tag{1-3}$$

One of Maxwell's theoretical findings was that the magnitudes of the electric field intensity and of the magnetic flux density in an electromagnetic wave are proportional to each other, with the velocity of light c as the proportionality constant. That is,

$$E = cB \tag{1-4}$$

in all electromagnetic waves. We are therefore able to rewrite Eq. (1-3) in the form

$$W = \frac{1}{2}\left(\epsilon_0 cEB + \frac{EB}{\mu_0 c}\right) A\,dx$$

$$= \frac{1}{2}\left(\epsilon_0 c + \frac{1}{\mu_0 c}\right) EBA\,dx. \tag{1-5}$$

Because

$$c = \frac{1}{\sqrt{\epsilon_0 \mu_0}},$$ (1–1)

we see that

$$\epsilon_0 c = \frac{1}{\mu_0 c},$$

and so

$$\left(\epsilon_0 c + \frac{1}{\mu_0 c} \right) = \frac{2}{\mu_0 c}.$$

Therefore Eq. (1–5) becomes

$$W = \frac{EBA\,dx}{\mu_0 c}.$$ (1–6)

The amount of energy W contained in the box at any instant is traveling at the speed c, and in the time $dt = dx/c$ it will all have passed out of the box. Thus the rate at which energy passes through the box is

$$P = \frac{dW}{dt} = \frac{Wc}{dx} = \frac{EBA}{\mu_0}.$$

The magnitude S of the Poynting vector is the rate of energy flow per unit cross-sectional area. Here the area of the box is A, and we have

$$S = \frac{P}{A} = \frac{EB}{\mu_0}.$$ (1–7)

Since $\mathbf{S}$ represents the flow of energy per unit area, it is expressed in watts/m².

In an actual electromagnetic wave both E and B vary in magnitude as shown in Fig. 1–4. While the average values of E and B individually are evidently zero, since each is negative as much as it is positive, the average value of the product EB is *not* zero. When E is negative B is negative, and when E is positive B is positive; their product is therefore always a positive quantity. The average value of EB, which we write $\overline{EB}$, is related to the maximum values of E and B by the formula

$$\overline{EB} = \tfrac{1}{2} E_{max} B_{max}.$$ (1–8)

Hence the average magnitude $\overline{S}$ of the Poynting vector is

$$\overline{S} = \frac{E_{max} B_{max}}{2\mu_0}.$$ (1–9)

It is also true that the average values of E^2 and B^2 are related to their maximum values by

$$\overline{E^2} = \tfrac{1}{2} E_{max}^2, \qquad \overline{B^2} = \tfrac{1}{2} B_{max}^2.$$ (1–10)

We can therefore express the average magnitude of the Poynting vector in the alternative forms

$$\overline{S} = \frac{cB_{\text{max}}^2}{2\mu_0} = \frac{E_{\text{max}}^2}{2c\mu_0} \qquad (1\text{–}11)$$

in view of Eq. (1–4).

Electromagnetic radiation from the sun arrives at the earth at the rate of about 1400 watts/m². A single electromagnetic wave of the kind we have been discussing (as distinct from the complex of different individual waves that comes from the sun) whose Poynting vector has this magnitude has a maximum electric field intensity of

$$E_{\text{max}} = \sqrt{2c\mu_0\overline{S}}$$

$$= \sqrt{2 \times 3 \times 10^8 \text{ m/sec} \times 1.26 \times 10^{-6} \text{ w/amp·m} \times 1400 \text{ watts/m}}$$

$$= 1.03 \times 10^3 \text{ volts/m}.$$

The corresponding maximum magnetic flux density is

$$B_{\text{max}} = \frac{E_{\text{max}}}{c} = \frac{1.03 \times 10^3 \text{ volts/m}}{3 \times 10^8 \text{ m/sec}}$$

$$= 3.43 \times 10^{-6} \frac{\text{w}}{\text{m}^2}.$$

1–5 RADIATION PRESSURE

Electromagnetic waves transport momentum as well as energy. The simplest method for demonstrating this makes use of the mass-energy relationship

$$E = mc^2.$$

A portion of an electromagnetic wave whose energy content is E is equivalent to a mass of

$$m = \frac{E}{c^2}.$$

Since the wave travels at the speed c, the momentum of this portion is

$$mc = \frac{E}{c}.$$

Hence the *momentum density* (momentum per unit volume) of an electromagnetic wave whose energy density is w is w/c, and the rate of flow of momentum per unit cross-sectional area of the wave is S/c, where $\mathbf{S}$ is the Poynting vector of the wave.

The most interesting consequence of momentum transport by electromagnetic waves is the phenomenon of *radiation pressure*, the pressure these waves exert on any surface they fall upon. The force a stream of particles exerts on a wall is equal to the rate of change of their momenta upon striking the wall. An electromagnetic wave whose Poynting vector has the magnitude S loses the momentum S/c per unit area per unit time when it is absorbed by a surface, and so the force it exerts upon the wall is S/c per unit area. Since pressure is force per unit area, the pressure p of the wave is

$$p = \frac{S}{c} \quad \text{(absorbed)} \tag{1–12}$$

as it is absorbed. If the wave is totally reflected, its change in momentum per unit area per unit time is doubled to $2S/c$ (since momentum is a vector quantity), and the pressure it exerts is also doubled to

$$p = \frac{2S}{c} \quad \text{(reflected)} . \tag{1–13}$$

As we noted in the previous section, the approximate magnitude of the Poynting vector for solar radiation at the earth is $S = 1400$ watts/m². If this radiation is totally absorbed, the resulting pressure on the absorber is

$$p = \frac{S}{c} = \frac{1400 \text{ watts/m}^2}{3 \times 10^8 \text{ m/sec}} = 4.7 \times 10^{-6} \text{ n/m}^2 .$$

The seemingly trivial pressure of solar radiation is responsible for the deflection of comet tails so that they always point away from the sun; it has even been seriously suggested as a means of interplanetary propulsion through the use of large reflecting "sails" on space ships.

If we confine a certain amount of electromagnetic radiation in a container with perfectly reflecting walls, the various waves will bounce around until their directions are random and their average density is uniform. To find the pressure this radiation exerts on the container walls we can proceed almost exactly as we do in the case of randomly moving molecules in the kinetic theory of gases (Appendix A). There we consider an assembly of N molecules, each with the average kinetic energy $\overline{\text{KE}}$, in a box of volume V. The pressure these molecules exert on the walls of the box is given by

$$p = \frac{2}{3} \frac{N}{V} \overline{\text{KE}},$$

which we can rewrite in terms of the energy density w of the molecules as

$$p = \tfrac{2}{3}w \quad \text{(molecules in a box)} \tag{1–14}$$

since

$$w = \frac{\text{total molecular kinetic energy}}{\text{volume}} = \frac{N \overline{\text{KE}}}{V} .$$

Equation (1–14) does not hold for trapped radiation because radiation possesses only half as much momentum for a given energy content as molecules do. To verify this statement we note that from the relation $mc = E/c$, the momentum of an electromagnetic wave of energy content E is

$$\text{Momentum} = \frac{E}{\text{speed}} \quad \text{(radiation)},$$

while the momentum of a molecule whose energy is KE is

$$\text{Momentum} = \frac{2\,\text{KE}}{\text{speed}} \quad \text{(molecules)}.$$

since its momentum is mv and its kinetic energy $\frac{1}{2}mv^2$. Because pressure results from momentum transfer and radiation of energy density w contains half the momentum of a group of molecules of the same energy density,

$$p = \tfrac{1}{2} \times \tfrac{2}{3}w = \tfrac{1}{3}w \quad \text{(radiation in a box).} \tag{1–15}$$

The pressure exerted by electromagnetic radiation in a container with reflecting walls is equal to one-third of its energy density.

1–6 BLACKBODY RADIATION

All objects give off electromagnetic radiation at a rate that increases with their temperature. Furthermore, the predominant wavelength in the radiation decreases with increasing temperature as an object is heated; the sequence of colors that appears starts with dull red and goes through bright red, orange, yellow, and white to bluish-white. Although a solid must have a temperature of at least 700°C or thereabouts to emit visible light of an intensity sufficient for it to be seen, all bodies at all temperatures radiate electromagnetic waves that can be detected by suitable techniques. The ability of a body to emit radiation is directly proportional to its ability to absorb radiation: a good absorber is a good emitter, and vice versa. This conclusion follows from the fact that a body at the same temperature as its environment must be absorbing and emitting radiation at exactly the same rates. When the temperature of the body exceeds that of its environment, it emits more radiation than it absorbs, and it is this difference we perceive.

A perfect absorber is one that absorbs all electromagnetic energy that falls upon it, regardless of wavelength. Such an absorber, called a *blackbody* by physicists even though actual objects painted black reflect at least a little radiation, is therefore also the best possible radiator at any temperature. It is convenient to discuss thermal radiation in terms of the behavior of an ideal blackbody in order to avoid such extraneous considerations as the nature of the emitting surface. A good approximation

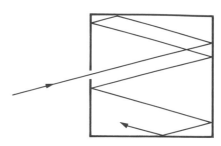

FIG. 1–7.   A small hole in a hollow object is
a good approximation of a blackbody.

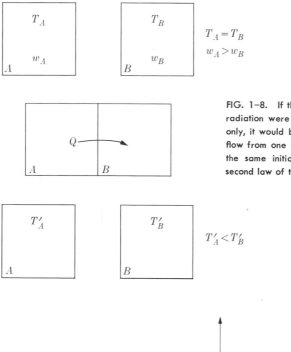

$$T_A = T_B$$
$$w_A > w_B$$

FIG. 1–8.   If the energy density w of blackbody
radiation were not a function of temperature T
only, it would be possible for radiant energy to
flow from one body to another when both have
the same initial temperature; this violates the
second law of thermodynamics.

$$T'_A < T'_B$$

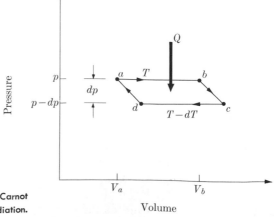

FIG. 1–9.   The operating cycle of a Carnot
engine whose working substance is radiation.

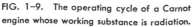

of a blackbody is a small hole in a hollow object (Fig. 1–7), since radiation striking the hole enters the cavity and is trapped there by multiple reflections even if the cavity walls are not perfectly absorbing.

A valuable theoretical approach to blackbody radiation is to consider a blackbody as a thermodynamic system whose working substance is electromagnetic radiation. We can use this approach to show that the energy density of such radiation within a cavity depends only upon the temperature of its walls. Let us suppose that this is *not* true. We imagine two cavities, A and B, whose walls are at the same temperature but cavity A has a higher energy density than cavity B (Fig. 1–8). We join the cavities together, and radiant energy proceeds to flow from A to B until their respective energy densities are equal. Then we separate the cavities and reseal them. Heat now flows from the walls of cavity A to the radiation remaining within to make up for the radiation lost to cavity B, the net result being a lower temperature for cavity A. At the same time some of the additional radiant energy in B is absorbed by its walls, the net result being a higher temperature for cavity B. We may next connect a heat engine between A and B, using the latter as the high-temperature reservoir and the former as the low-temperature one. After the excess heat in A has been converted to work, we can reconnect the cavities, which are again at the same temperature, and then repeat the sequence until no energy is left in both cavities. But the second law of thermodynamics states that it is impossible to construct an engine that, operating in a cycle, produces no effect other than the extraction of heat from a reservoir and the performance of an equivalent amount of work—and the two cavities, initially at the same temperature, constitute a single reservoir from the point of view of thermodynamics. Hence we conclude that the energy density of the radiation within a cavity depends solely upon the cavity temperature. Since a blackbody is equivalent to a hole in the wall of a cavity, it follows that the rate at which a blackbody radiates energy is likewise a function solely of the body's temperature.

We can use another thermodynamic argument to find precisely how the rate at which a blackbody radiates energy varies with its temperature. Let us consider a cylinder whose walls are perfectly reflecting which has a movable piston, also a perfect reflector, at one end. The cylinder contains electromagnetic radiation, and we can regard it as a Carnot engine whose working substance is the radiation. Figure 1–9 shows the operating cycle of this engine. The first stage is an isothermal expansion at the temperature T, during which the cylinder volume increases from V_a to V_b. Because the pressure of the radiation is proportional to the energy density, which in turn depends only upon temperature, p remains constant during this process. The radiation is next allowed to expand adiabatically from b to c, and in so doing its temperature drops by dT to $T - dT$ and its pressure drops by dp to $p - dp$. Then an isothermal compression from c to d takes place, followed by an adiabatic compression to a, the starting point.

The work dW done by the radiation in expanding isothermally by dV is, since $p = \frac{1}{3}w$,

$$dW = p\,dV = \frac{w}{3}\,dV.$$

The internal energy of the radiation is the product of the energy density w and the volume V. Hence the increase dU in internal energy during an isothermal expansion of dV is

$$dU = w\,dV.$$

The heat input Q in the expansion from V_a to V_b is, from the first law of thermodynamics,

$$Q = \int_a^b dQ = \int_a^b dU + \int_a^b dW$$

$$= \int_{V_a}^{V_b} w\,dV + \int_{V_a}^{V_b} \frac{w}{3}\,dV$$

$$= \tfrac{4}{3}w(V_b - V_a).$$

The work dW done during the entire cycle is the area of the parallelogram $abcd$,

$$dW = dp(V_b - V_a) = \frac{dw}{3}\,(V_b - V_a).$$

The efficiency Eff of all Carnot engines is

$$\text{Eff} = \frac{dW}{Q} = 1 - \frac{T_2}{T_1}.$$

Here

$$1 - \frac{T_2}{T_1} = 1 - \frac{T - dT}{T} = \frac{dT}{T}$$

and

$$\frac{dW}{Q} = \frac{dw}{4w}.$$

Hence

$$\frac{dw}{4w} = \frac{dT}{T}$$

and

$$\int \frac{dw}{w} = 4\int \frac{dT}{T}.$$

If we let $\ln b$ represent the constant of integration, we obtain

$$\ln w = 4 \ln T + \ln b,$$

$$w = bT^4. \tag{1–16}$$

This result states that the energy density of the radiation trapped in a container with reflecting walls is proportional to T^4, the fourth power of its absolute temperature. If we bore a hole in the container, radiation will leak out at a rate proportional to w and hence to T^4. Since such a hole is in effect a blackbody, the amount of energy R radiated by a blackbody per unit time and per unit area is also proportional to T^4:

$$R = \sigma T^4. \tag{1–17}$$

Equation (1–17) is known as the *Stefan-Boltzmann law* because it was arrived at empirically by Josef Stefan in 1879 and, five years later, was theoretically derived by Ludwig Boltzmann in the above way. The value of the constant σ is

$$\sigma \doteq 5.67 \times 10^{-8} \text{ watt/m}^2 \cdot {}^\circ\text{K}^4.$$

Thus a blackbody whose area is A and whose temperature is T radiates electromagnetic energy at the rate of $\sigma A T^4$ j/sec. We note that Eq. (1–17) tells us only how much energy is emitted, and nothing about the manner in which the energy is divided among the infinite number of possible wavelengths.

1–7 THE RAYLEIGH-JEANS FORMULA

An argument almost identical to the one we used above to show that the total radiation rate from a blackbody must depend solely upon temperature may be used to show that the distribution of energy among the various wavelengths of the radiation from a blackbody must also depend solely upon its temperature. In this case we employ a filter between the cavities when they are joined together that permits radiation in a very narrow wavelength interval only to pass from one to the other, thus restricting the hypothetical energy transfer to that wavelength interval. The rate at which energy is radiated in this interval is the same for all blackbodies at the same temperature if the second law of thermodynamics is not to be violated.

The empirical data on the spectra of blackbody radiation at various temperatures are shown in Fig. 1–10. The ordinate of the graph is the monochromatic energy density w_λ, where $w_\lambda \, d\lambda$ is the amount of radiant energy per unit volume with wavelengths between λ and $\lambda + d\lambda$. We note that at high temperatures the areas under the curves, which equal the total energy in the radiation, are greater than at low temperatures, and the positions of the maxima in the curves, which represent the pre-

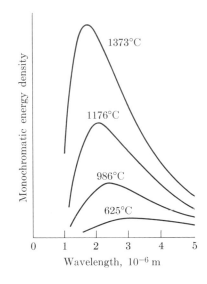

FIG. 1–10. The monochromatic energy density of blackbody radiation as a function of temperature.

dominant color, shift to shorter wavelengths as T increases. Both of these effects are entirely in accord with classical physics. As we saw in the previous section, the area under each curve is given by

$$\int_0^\infty w_\lambda \, d\lambda = w = bT^4.$$

The wavelength shift, too, can be derived by thermodynamic reasoning. In agreement with experiment, the wavelength λ_{max} at which w_λ is a maximum turns out to be inversely proportional to the absolute temperature of the radiating blackbody. That is,

$$\lambda_{max} = \frac{C}{T}, \qquad \lambda_{max} T = C, \tag{1–18}$$

where $C = 2.898 \times 10^{-3}$ m·°K. This relationship is called the *Wien displacement law*.

Still, despite these successes, we have no formula that gives the distribution of wavelengths in the blackbody spectrum at a given temperature. The problem of deriving such a formula was tackled at the turn of the last century by Lord Rayleigh and James Jeans, who employed an ingenious but rigorous approach. They considered the radiation within a cavity of temperature T whose walls are perfectly reflecting as a series of standing electromagnetic waves (Fig. 1–11), essentially a three-dimensional generalization of standing waves in a stretched string. The condition for standing waves in this case is that the path length from wall to wall, in any direction, be an integral multiple of a half-wavelength; otherwise the waves would have to penetrate the walls, which contradicts the

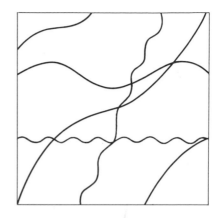

FIG. 1–11. Several of the possible standing waves that can exist in a cavity.

premise of perfectly reflecting walls. Each standing wave represents a mode of vibration, of which there are obviously a great many owing to the multiplicity of possible paths. Rayleigh and Jeans first calculated the number of modes of vibration per unit volume dn_λ within the cavity whose wavelengths lie between λ and $\lambda + d\lambda$, which they found to be

$$dn_\lambda = \frac{8\pi\, d\lambda}{\lambda^4}. \tag{1–19}$$

Next, Rayleigh and Jeans drew upon the theorem of equipartition of energy, which, as we know, states that if a system can possess energy in a number of independent ways, called degrees of freedom, each degree of freedom has an average energy of $\frac{1}{2}kT$. Each mode of vibration within a radiation-filled cavity constitutes *two* degrees of freedom. We can arrive at this conclusion in two ways. Each mode of vibration must be caused by an atomic oscillator in the cavity wall, and such an oscillator has two degrees of freedom, one representing its kinetic energy and one its potential energy. Alternatively we may note that every electromagnetic wave has electric energy in its electric field and magnetic energy in its magnetic field, again adding up to two modes of energy possession and hence two degrees of freedom. Therefore each mode of vibration has the average energy $2 \times \frac{1}{2}kT = kT$. Since there are

$$dn_\lambda = \frac{8\pi\, d\lambda}{\lambda^4}$$

modes of vibration per unit volume in the cavity in the wavelength interval between λ and $\lambda + d\lambda$ and each has the average energy kT, the total energy per unit volume contained in the radiation within this wavelength interval is

$$E_\lambda\, d\lambda = kT\, dn_\lambda = \frac{8\pi kT\, d\lambda}{\lambda^4}. \tag{1–20}$$

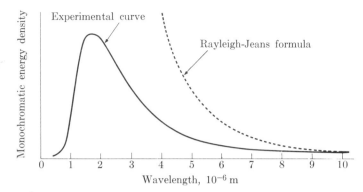

FIG. 1-12. Comparison of the Rayleigh-Jeans formula with the experimental data for blackbody radiation at a temperature of 1600° K.

Equation (1–20) is known as the *Rayleigh-Jeans formula*. To express this formula in terms of frequency f rather than wavelength, we note that

$$f = c/\lambda, \qquad df = -c\, d\lambda/\lambda^2, \qquad E_f\, df = -E_\lambda\, d\lambda,$$

since an increase in frequency corresponds to a decrease in wavelength. Hence

$$E_f\, df = \frac{8\pi f^2 kT\, df}{c^3}. \tag{1–21}$$

Even a casual inspection of the Rayleigh-Jeans formula, which contains everything that classical physics can say about the spectrum of blackbody radiation, reveals that it cannot possibly be correct. As λ decreases, corresponding to the ultraviolet end of the spectrum, Eq. (1–20) predicts an increase in the radiated energy at a rapidly (as $1/\lambda^4$) increasing rate. In fact, in the limit of infinitely short wavelength, $E\, d\lambda$ approaches infinity—the "ultraviolet catastrophe" of classical physics, since, of course, $E\, d\lambda$ actually goes to zero as λ approaches zero. According to the Rayleigh-Jeans formula, the total energy density in a cavity at *any* temperature, however low, is

$$E = \int_0^\infty E_\lambda\, d\lambda = \infty,$$

which is nonsense, instead of the correct Stefan-Boltzmann result of bT^4.

Oddly enough, considering its hopeless predictions for the short-wavelength end of the radiation spectrum, the Rayleigh-Jeans formula agrees perfectly with observation at long wavelengths. This is evident in Fig. 1–12, where the dashed line was calculated from Eq. (1–20). Because the Rayleigh-Jeans formula contains no arbitrary constants, this partial agreement indicates that it must have some germs of truth in it. The question is, which of the classical laws that went into the Rayleigh-Jeans formula are not really correct? We shall learn in Chapter 3 just how radical a change in the conceptual structure of physics' was required in order to account for the observed properties of electromagnetic radiation.

PROBLEMS

1. A radio transmitter has a vertical antenna. Does it matter whether the receiving antenna is vertical or horizontal? Explain.

2. At a particular instant the magnetic flux density in an electromagnetic wave is 5×10^{-7} w/m². Find the electric field intensity at that instant.

3. At a particular instant the electric field intensity in an electromagnetic wave is 3×10^{-3} volts/m. Find the magnetic induction at that instant.

4. Find the pressure exerted by a radio wave which transmits energy at the rate of 1 watt/m² when it is reflected by a surface.

5. An electromagnetic wave whose Poynting vector is 12 watts/m² is absorbed by a surface whose area is 6 m². Find the force on the surface.

6. A small marine radar operates at a wavelength of 3.2 cm and a power of 7 kw per pulse. (a) What is the frequency of the radar waves? (b) How much reaction force is exerted on the antenna when each pulse is emitted?

7. The intensity of solar radiation varies inversely with the square of the distance from the sun. The earth is an average of 1.5×10^8 km from the sun. How close to the sun is it necessary to be for its radiation pressure to be 1 n/m²?

8. A radio wave is found to exert a pressure of 10^{-8} n/m² on a reflecting surface. (a) How much power per unit area does the wave transport? (b) What are the maximum magnitudes of its electric intensity and magnetic flux density?

9. For good reception a radio wave should have a maximum electric field intensity of at least 10^{-4} volt/m when it arrives at the receiving antenna. (a) What is the maximum flux density of the magnetic field of such a wave? (b) What is the magnitude of the Poynting vector of such a wave? (c) What radiation pressure does it exert when it is absorbed?

10. The entire 1000-watt luminous output of a searchlight is focused on a spot 2 m in diameter. If half of the light is absorbed and half reflected, how much pressure does it exert?

11. Electromagnetic radiation carries both energy and momentum. Why does the momentum of the sun not decrease with time as its energy content does?

12. Suppose an aircraft is being designed which is to be supported against gravity by radiation pressure from a searchlight on the ground. The lower surface of the aircraft is a perfect reflector, and the entire luminous output of the searchlight is focused on it. If the aircraft's mass is 100 kg, find the required power in watts of the searchlight. Neglecting practical considerations such as aiming the beam, does the horizontal area of the aircraft matter?

13. A *laser* is a device that produces a very intense beam of light that is mono-chromatic (all one wavelength), coherent (all the individual waves are in step with one another, which increases their effectiveness), and practically nondivergent (the beam stays parallel without spreading out). Find the pressure exerted on an absorbing surface by a laser beam in which $E_{max} = 10^{10}$ volts/m. How many atm is this?

2

Relativity

"Turning to the theory of relativity itself, I am anxious to draw attention to the fact that this theory is not speculative in origin; it owes its invention entirely to the desire to make physical theory fit observed fact as well as possible. We have here no revolutionary act but the natural continuation of a line that can be traced through centuries. The abandonment of a certain concept connected with space, time, and motion hitherto treated as fundamental must not be regarded as arbitrary, but only as conditioned by observed facts. The law of the constant velocity of light in empty space, which has been confirmed by the development of electro-dynamics and optics, and the equal legitimacy of all inertial systems (special principle of relativity), which was proved in a particularly incisive manner by Michelson's famous experiment, between them made it necessary, to begin with, that the concept of time should be made relative, each inertial system being given its own special time. As this notion was developed it became clear that the connection between immediate experience on one side and coordinates and time on the other had hitherto not been thought out with sufficient precision. It is in general one of the essential features of the theory of relativity that it is at pains to work out the relations between general concepts and empirical facts more precisely. The fundamental principle here is that the justification for a physical concept lies exclusively in its clear and unambiguous relation to facts that can be experienced. According to the special theory of relativity, spatial coordinates and time still have an absolute character in so far as they are directly measurable by stationary clocks and bodies. But they are relative in so far as they depend on the state of motion of the selected inertial system. According to the special theory of relativity the four-dimensional continuum formed by the union of space and time retains the absolute character which, according to the earlier theory, belonged both to space and time separately. The influence of motion (relative to the coordinate system) on the form of bodies and on the motion of clocks, also the equivalence of energy and inert mass, follow from the interpretation of coordinates and time as products of measurement."

Albert Einstein (1879–1955)

The great French mathematician and physicist J. L. Lagrange (1736–1813) once wrote that "Newton was the greatest genius who ever lived, and the most fortunate, for there cannot be more than once a system of the world to establish." Though Newton has lost none of his glory since this was written, our own century has seen the basis for the second part of Lagrange's tribute disappear. Not only have relativity and quantum theory forced major revisions in our ideas of the "system of the world," but they have also shown the folly in assuming that we have complete understanding of any major aspect of the physical universe. Today's measurements may yield no discrepancy with a particular theory, but we have no way of knowing what the more precise or more searching measurements of the future will have to say. Science is not like mountain climbing, where there is a definite goal to reach and either success or failure; on the contrary, it is a progressive endeavor, with further heights coming into view each time we think we have, at long last, achieved a summit.

2–1 THE MICHELSON–MORLEY EXPERIMENT

When we speak of light waves traveling through space, precisely what do we mean? If space is empty, there is nothing to vibrate back and forth, and so the concept of a wave moving through empty space is not easy to accept. In the 19th century, physicists avoided the problem by assuming that the universe is pervaded with a curious fluid called *ether* whose only property is an ability to support electromagnetic waves. This seems a harmless enough idea, but it contains an implication that led directly to its downfall and thereby to the theory of relativity. Let us see what this implication is through a simple analogy.

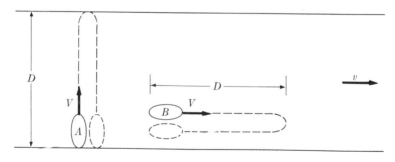

FIG. 2–1

Figure 2–1 is a sketch of a river flowing with the speed v. The river's width is D. Two identical boats start out from one bank of the river with the same speed V; boat A is to cross the river to a point on the other bank directly opposite the starting place and then to return, while boat B is to proceed downstream for the distance D and then to return to the starting place. How much time will each round trip take?

We shall first consider boat A. If A heads directly across the river (as in Fig. 2–2a), the current will carry it downstream from its destination on the other bank opposite where it started, and so it must head somewhat upstream (Fig. 2–2b) in order to compensate for the current. When it does this, it will have an upstream component of its velocity

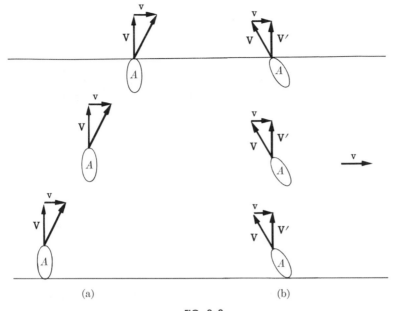

(a)   (b)

FIG. 2-2

equal to exactly $-v$ in order to cancel out the downstream current v, leaving the component V' as its net speed across the river. From Fig. 2-2 we see that these speeds are related by

$$V^2 = V'^2 + v^2,$$

so that the actual speed with which it crosses the river is

$$V' = \sqrt{V^2 - v^2} = V\sqrt{1 - (v^2/v^2)}.$$

Hence the time for the crossing is the distance D divided by the speed V', and, since the reverse crossing involves exactly the same amount of time, the total round-trip time t_A is twice D/V', or

$$t_A = \frac{2D/V}{\sqrt{1 - (v^2/V^2)}}. \qquad (2\text{-}1)$$

Boat B is in a different situation. As it proceeds downstream, its speed relative to the shore is its own speed V *plus* the speed v of the river (Fig. 2-3), and it travels the distance D downstream in the time

$$\frac{D}{V + v}.$$

On the return trip, however, B's speed relative to the shore is its own speed V *minus* the speed v of the river, which now opposes its motion,

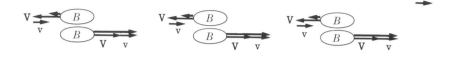

FIG. 2-3

and it requires the longer time

$$\frac{D}{V - v}$$

to travel upstream the distance D. The total round trip time t_B is the sum of these two, or

$$t_B = \frac{D}{V + v} + \frac{D}{V - v}.$$

Using the common denominator $(V + v)\,(V - v)$ for both terms, we have

$$t_B = \frac{D(V - v) + D(V + v)}{(V + v)(V - v)}$$

$$= \frac{2DV}{V^2 - v^2}$$

$$= \frac{2D/V}{1 - (v^2/V^2)}, \tag{2-2}$$

which is longer than t_A. The ratio between the times t_A and t_B is

$$\frac{t_A}{t_B} = \sqrt{1 - (v^2/V^2)}. \tag{2-3}$$

By sending out a pair of boats in this way, then, a measurement of t_A/t_B together with a knowledge of the boats' speed V enables us to calculate the speed v of the river current.

Now let us restate the above problem in terms of the passage of light waves through the ether. Instead of a current of water relative to the shore we shall have a current of ether relative to the earth as the earth moves around the sun within the conjectured universal sea of ether, and instead of a pair of boats we shall have a pair of light beams from a single source. Figure 2–4 is a schematic representation of the apparatus used in the famous experiment performed by the American physicists Michelson and Morley in 1887. As shown in the figure, parallel light strikes the half-silvered mirror A and is split into two beams, one proceeding in the original direction to the right and the other reflected up-

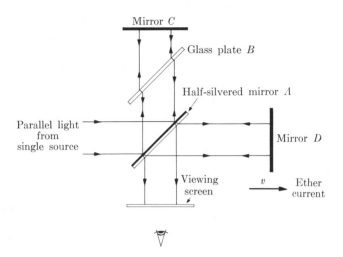

FIG. 2–4. Schematic diagram of the apparatus used by Michelson and Morley to measure the speed of the ether current past the earth.

ward. The former beam is reflected by the mirror D and returns to A, where some of it is further reflected to a viewing screen. The upward beam is reflected by the mirror C and also returns to A, where part of it is transmitted to reach the viewing screen. The distances from A to C and from A to D are the same, and a glass plate B is inserted in the path of the upward beam, so that each beam passes through identical thicknesses of air and glass between its entry into the apparatus and its arrival at the viewing screen. The presence of an "ether current" as shown would result in the different beams requiring different transit times to go from the source to the viewing screen. Instead of arriving at exactly the same stage in their cycles, the light waves of the two beams would be slightly—but definitely—out of step with each other, and hence would interfere destructively to a corresponding extent.

Michelson and Morley expected to find a value for t_A/t_B of **0.999999995**, corresponding to the substitution in Eq. (2–3) of the speed of light c (186,000 mi/sec) for the boats' speed V and of the earth's orbital speed (averaging 18.5 mi/sec) for the river current v, namely

$$\frac{t_A}{t_B} = \sqrt{1 - (v^2/c^2)}. \qquad (2\text{–}4)$$

The results of the Michelson-Morley experiment were startling: although their apparatus was capable of measuring the minute time difference expected, *none was found*. When the experiment was performed at different seasons and in different locations, and when different experiments were tried for the same purpose, the conclusions were always identical—no motion through the ether could be detected. To give a somewhat fanciful analogy, it was as if our boats made their round trips in a river bed running down the side of a mountain and returned at the same time: if there had been water in the river bed, it would have been in the form of a strong current; if there had been a strong current, the travel times would have differed; they did not differ, hence the river bed must have been empty. (For this analogy to hold we assume the boats were also equipped with wheels, permitting them to travel with equal facility on dry land and in water.)

2–2 FRAMES OF REFERENCE

The immediate conclusion we can draw from the failure of the Michelson-Morley experiment to detect any effect of the ether on the motion of light waves is that the speed of light in free space is a constant everywhere, completely independent of any relative motion of the ether. As a corollary we must discard the notion of the ether entirely; the sole property of the ether is its ability to support electromagnetic waves, something it cannot be doing if such waves are not affected by its motion past source and observer.

The absence of an ether has a profound consequence that is not evident from our discussion thus far. Let us consider for a moment what we mean by so elementary a notion as velocity. When we say that a body is moving we imply the existence of some sort of frame of reference (coordinate system) in which its velocity v is to be measured. Suppose that a car traveling down a road goes past a slower car. There is a thick fog, so neither driver can see anything but the road and the other car. To the driver in the fast car, the other car appears to be moving backward. The driver in the slow car, while observing that the fast car is going past him, nevertheless is sure that he is moving in the same direction as it is.

Who is correct? The modern answer is that both are correct in terms of their own frames of reference, with each driver determining his own motion relative to the ground while determining that of the other car relative to himself. Of course, we tend to think offhand that the driver of the slow car is "correct" because the earth is the frame of reference for our daily activities. But the earth itself is rotating on its axis, revolving about the sun, and traveling with the sun as the sun revolves about the center of our galaxy and partakes with the galaxy of the latter's motion through space. How then can we separate "apparent" from "true" motions, and establish the "truth" about what is happening in the universe?

At one time the ether looked like the solution to the problem of finding the "true" frame of reference of the universe. Newton's laws were supposed to hold exactly when phenomena are viewed from a frame of reference stationary relative to the ether. Now there would be no ambiguity in physical laws, all of which could be referred to the ether. However, as the Michelson-Morley experiment revealed, there is *no* ether, and therefore no way of specifying any universal frame of reference. In other words, if we observe something changing its position with respect to us, we have no way (even in principle) of knowing whether *it* is moving or *we* are moving. If we were alone in the universe we would have no way of finding out whether we are moving or not. All motion is relative to the observer; there is no such thing as absolute motion.

2–3 THE SPECIAL THEORY OF RELATIVITY

The theory of relativity is a largely successful attempt to work out the physical consequences of the absence of a universal frame of reference. The special theory of relativity, published in 1905 by Albert Einstein, confines itself to problems involving the motion of frames of reference at constant velocity (that is, both constant speed and constant direction) with respect to one another; the general theory of relativity, published 10 years later by Einstein, deals with problems involving frames of reference accelerated with respect to one another. The special theory has had an enormous impact on all of physics, and since it is also simpler mathematically than the general theory, we shall consider it exclusively.

Einstein developed the special theory of relativity from two postulates. The first states that

> *The laws of physics may be expressed in the same set of equations for all frames of reference moving at constant velocity with respect to one another.*

This postulate follows directly from the absence of a universal frame of reference. If the laws of physics took on different forms for different observers in relative motion, they could infer from these differences which of them were "stationary" in space and which were "moving"; but this distinction does not exist in nature, since there is no universal frame of reference, and the above postulate is an expression of this fact.

The second postulate states that

> *The speed of light in free space has the same value for all observers, regardless of their state of motion.*

This postulate expresses the result of the Michelson-Morley experiment. At first glance these postulates do not seem so very extraordinary, a misleading impression indeed, as we can show with a simple example. Suppose that we have the two boats *A* and *B* once more, this time in a

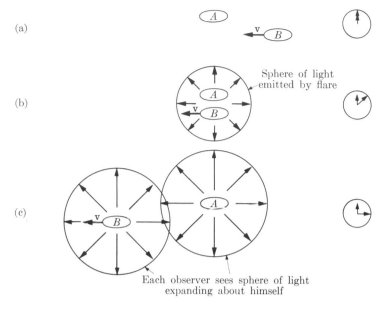

(a)

(b)

Sphere of light
emitted by flare

(c)

Each observer sees sphere of light
expanding about himself

FIG. 2-5

large lake (Fig. 2-5a). Boat A is stationary and boat B is drifting at
the constant velocity v. There is a fog present, and the observers in each
boat have no idea which boat is moving. At the instant that B is abreast
of A (Fig. 2-5b), somebody in one of the boats fires a flare. That light
from the flare spreads out uniformly in all directions, something we could
verify if our visual responses were fast enough by watching the light scattered by the fog. An observer on either boat must see a sphere of light expanding with *him* at its center (Fig. 2-5c), according to the postulates of
special relativity, even though one of them is changing his position with
respect to the point where the flare went off. But it is impossible for the
observers to tell which of them is undergoing such a change in position,
since (owing in this case to the fog) there is no frame of reference other
than each boat itself, and so, since the speed of light is the same for
both of them, they must both see the identical phenomenon. And, of
course, the fog is irrelevant; nothing would be changed in its absence
except our ability to visualize what is going on.

What is so remarkable about the situation of Fig. 2-5? Let us consider a more familiar analog. It is a clear day and, instead of firing a
flare, somebody drops a stone in the water. A circular pattern of ripples
spreads out, as in Fig. 2-6, *which appears different* to observers on each
boat. Merely by observing whether or not they are at the center of the
pattern of ripples, the observers can tell whether they are stationary
with respect to the water or not. Here the water acts as a frame of

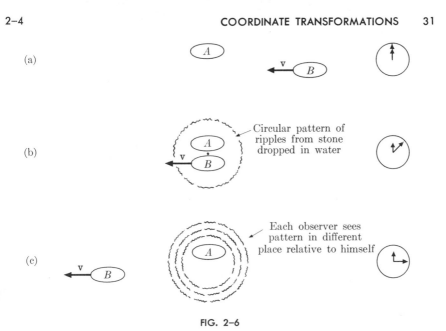

(a)

(b)

Circular pattern of
ripples from stone
dropped in water

(c)

Each observer sees
pattern in different
place relative to himself

FIG. 2–6

reference, and an observer on a boat moving through the water will measure a ripple speed with respect to him different from that measured by an observer on a stationary boat. It is important for us to note that motion and waves *in water* are entirely different from motion and waves *in space;* water is in itself a frame of reference while space is not, and wave speeds in water vary with the observer's motion while wave speeds in space do not.

There is only one way of accounting for the fact that observers in the two boats observe identical expanding spheres of light, and that is to assume that the coordinate system of each observer, as seen by the other, is affected by their relative motion. When this idea is worked out, using no more than the ordinary laws of physics and Einstein's postulates, it is found that a host of peculiar effects arise—and all of them, no matter how strange to us, have been confirmed by experiment.

2–4 COORDINATE TRANSFORMATIONS

An event of some kind occurs at a certain moment somewhere and is witnessed by two observers. One of them is in a frame of reference S, and he finds that the event took place at the time t in the position specified by the coordinates x, y, z. The other observer is in a different frame of reference, S', which is identical with S except that it is moving at the constant velocity v in the $+x$-direction (Fig. 2–7). This observer finds that the same event took place at the time t' and has the coordinates x', y', z'. What is the relationship between these sets of measurements?

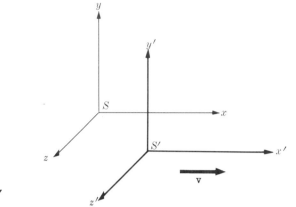

FIG. 2–7

Before relativity the question seemed a trivial one. Let us suppose that the clocks in both systems are started at the instant when the origins of S and S' coincided. If this is true, x measurements made in S at the time t exceed similar measurements made in S' by vt, since vt is the distance through which S' has moved relative to S in the x-direction during t. Hence $x = x' + vt$. In the absence of relative motion between S and S' in the y- and z-directions, $y' = y$ and $z' = z$. We assume that the clocks in S and S' are identical, and so there seems to be no reason why t' should not equal t. The set of equations

$$x' = x - vt, \tag{2–5}$$

$$y' = y, \tag{2–6}$$

$$z' = z, \tag{2–7}$$

$$t' = t \tag{2–8}$$

that relates measurements made in S' with those made in S is called the *Galilean transformation*. To obtain the corresponding transformation for velocity components measured in one system to those measured in the other, we merely differentiate the above expressions for x', y', and z' with respect to t'. This procedure yields

$$v'_x = \frac{dx'}{dt'} = v_x - v, \tag{2–9}$$

$$v'_y = \frac{dy'}{dt'} = v_y, \tag{2–10}$$

$$v'_z = \frac{dz'}{dt'} = v_z. \tag{2–11}$$

The above coordinate and velocity transformations agree completely with our everyday experience, but they do *not* agree with either of the postulates of special relativity. According to the first of these postulates, which is based on the absence of a universal "fixed" frame of reference, the formulas of physics must be the same in both S and S'. However, the basic equations of electricity and magnetism are different in S and S' when Eqs. (2–5) to (2–8) are used to transform measurements made in one frame to their equivalent values in the other, a fact that had caught the attention of physicists even before 1905 although its full significance was not then understood. The second postulate of relativity, which requires that the speed of light in free space be the same for all observers, is violated by the Galilean velocity transformation. If a light ray travels in the $+x$-direction at the speed $v_x = c$ in frame S, in the frame S' its speed is given by Eq. (2–9) as $v'_x = c - v$. While at first glance the Galilean transformation seems reasonable, it is inconsistent with the postulates of relativity and therefore cannot be correct.

Instead of relying upon our intuition, as we did in obtaining the Galilean transformation, we can derive a set of transformation equations directly from the postulates of special relativity. Because $x' = x - vt$ is in accord with ordinary mechanics, the new transformation for the x-coordinate must reduce to this formula when the relative velocity v is small. The simplest possibility is

$$x' = k(x - vt), \tag{2–12}$$

where k does not depend upon either of the coordinates x or t but may vary with v. This formula has the virtue of being linear in x and x', so that there is automatically one and only one value of x' for a given value of x without our having to impose any subsidiary conditions. From the first postulate of relativity, the inverse transformation from measurements made in S' to those made in S must have precisely the same form as Eq. (2–12) except for a change in the sign of v, and must also have the same constant of proportionality:

$$x = k(x' + vt). \tag{2–13}$$

As before, we have no reason to suppose that there is any difference between y and y' or between z and z', since the relative motion between S and S' is confined to the x-direction:

$$y' = y, \tag{2–14}$$
$$z' = z. \tag{2–15}$$

However, this equality does not extend to t and t', since Eqs. (2–12) and (2–13) cannot both be true if $t' = t$. If we substitute the value of x' given by Eq. (2–12) into Eq. (2–13) we obtain

$$x = k^2(x - vt) + kvt',$$

and so

$$t' = kt + \left(\frac{1 - k^2}{kv}\right)x. \tag{2–16}$$

The transformation specified by Eqs. (2–12) and (2–14) through (2–16) satisfies the first postulate of special relativity.

To determine the value of k we return to the situation pictured in Fig. 2–5 of a flare being set off at the instant when two boats in relative motion are directly abreast. Regarding the boats as frames of reference, we can choose their respective coordinates so that the relative velocity v is in the x-direction and, when the flare is fired, $x = x' = 0$ and $t = t' = 0$. An observer in each boat measures the speed with which the pulse of light from the flare spreads out into space. The second postulate of relativity requires that both observers determine the same speed of c, as we said earlier; thus the x-coordinate of the light pulse in the S-system obeys the formula

$$x = ct, \tag{2–17}$$

while in the S'-system the x'-coordinate of the pulse obeys the formula

$$x' = ct'. \tag{2–18}$$

We now use Eqs. (2–13) and (2–16) to convert the x'- and t'-measurements of Eq. (2–18) to their corresponding x- and t-values, which yields

$$k(x - vt) = ckt + \left(\frac{1 - k^2}{kv}\right)cx.$$

Solving for x, we find

$$x = ct\left[\frac{1 + v/c}{1 - (c/v)(1/k^2 - 1)}\right].$$

This formula for x must be the same as that of Eq. (2–17), namely $x = ct$. Equating these expressions for x, we have

$$ct = ct\left[\frac{1 + v/c}{1 - (c/v)(1/k^2 - 1)}\right]$$

from which we obtain

$$k = \frac{1}{\sqrt{1 - v^2/c^2}}. \tag{2–19}$$

This result is in accord with our expectations, since when the relative velocity v is small compared with the velocity of light c, $k \approx 1$ and Eqs. (2–13) and (2–16) reduce to their Galilean equivalents that we know are adequate in ordinary mechanics. Evidently "relativistic" effects can be expected only at extremely high speeds.

When the above equation for k is inserted in Eqs. (2–13) and (2–16) we obtain the *Lorentz transformation*, which relates measurements of the same event in the S- and S'-frames of reference:

$$x' = \frac{x - vt}{\sqrt{1 - v^2/c^2}}, \tag{2-20}$$

$$y' = y, \tag{2-21}$$

$$z' = z, \tag{2-22}$$

$$t' = \frac{t - vx/c^2}{\sqrt{1 - v^2/c^2}}. \tag{2-23}$$

This transformation is named in honor of H. A. Lorentz, a Dutch physicist, who proved that it is the only one that permits the laws of electricity and magnetism to have the same forms in frames of reference that are in relative motion. Einstein was able to show its more general significance as the proper coordinate transformation for all measurements, electromagnetic or not.

Equations (2–20) through (2–23) convert measurements of time and space made in the stationary S-frame to their counterparts in the moving S'-frame. As seen from frame S, frame S' is moving with the velocity v along the x-axis, while as seen from frame S', frame S is moving with the velocity $-v$ along the x-axis. Hence the *inverse Lorentz transformation*, which converts measurements of time and space made in the S'-frame to their counterparts in the S-frame, may be obtained from the Lorentz transformation by changing the sign of v. The inverse transformation therefore consists of the equations

$$x = \frac{x' + vt'}{\sqrt{1 - v^2/c^2}}, \tag{2-24}$$

$$y = y', \tag{2-25}$$

$$z = z', \tag{2-26}$$

$$t = \frac{t' + vx/c^2}{\sqrt{1 - v^2/c^2}}. \tag{2-27}$$

2–5 THE LORENTZ–FITZGERALD CONTRACTION

Measurements of space and time, as we have seen, are not absolute but depend upon the relative motion of the observer and that which is being observed. Different observers may obtain different results when they examine the same phenomenon, and all of them may be correct. This situation is a consequence of the fundamental character of the physical universe, and it should not really be surprising that the space-time

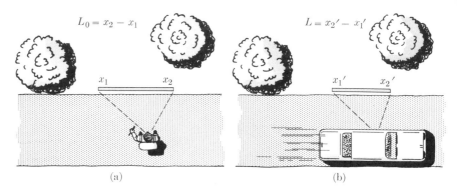

FIG. 2–8. (a) An observer at rest relative to a rod measures its length as L_0. (b) An observer moving relative to the same rod measures its length as L.

geometry of nature is not the same as the space-time geometry of Euclid and Galileo. The postulates of special relativity are perfectly reasonable, and if the deductions we are about to obtain from them seem strange, it is only because our personal experience involves frames of reference whose relative motion is on too trifling a scale to bring out relativistic effects.

Let us begin by considering a stick lying on the ground beside a road, as shown in Fig. 2–8. An observer at rest relative to the stick finds that one end is at $x = x_1$ and the other at $x = x_2$, so that its length L_0 is

$$L_0 = x_2 - x_1.$$

Now a car comes along and, as it passes the stick, a rider measures the length of the stick by determining the positions x'_1 and x'_2 of its ends in his frame of reference. To the rider, who is moving relative to the stick at the constant velocity v, the length of the stick is

$$L = x'_2 - x'_1.$$

How are L_0 and L related? From the inverse Lorentz transformation, Eq. (2–24), we have

$$x_1 = \frac{x'_1 + vt'}{\sqrt{1 - v^2/c^2}} \quad \text{and} \quad x_2 = \frac{x'_2 + vt'}{\sqrt{1 - v^2/c^2}},$$

so that

$$L_0 = x_2 - x_1 = \frac{x'_2 - x'_1}{\sqrt{1 - v^2/c^2}} = \frac{L}{\sqrt{1 - v^2/c^2}}$$

and

$$L = L_0\sqrt{1 - v^2/c^2}. \tag{2-28}$$

The length of an object in motion with respect to an observer is measured by the observer to be shorter than when it is at rest with respect to him. This shortening, known as the *Lorentz-Fitzgerald contraction*, works both ways; to a man in a rocket, measurements indicate that objects on the earth behind him are shorter than they were when he was on the ground by the amount specified in Eq. (2–28). The length of an object is a maximum when measured in a reference frame in which it is stationary, and its length is less when measured in a reference frame in which it is moving.

The relativistic length contraction is negligible for ordinary speeds, but is an important effect at speeds close to the speed of light. A speed of 1000 mi/sec seems enormous to us, yet it results in a shortening in the direction of motion to only

$$\frac{L}{L_0} = \sqrt{1 - (v^2/c^2)} = \sqrt{1 - \frac{(1000 \text{ mi/sec})^2}{(186,000 \text{ mi/sec})^2}}$$

$$= 0.999985 = 99.9985\%$$

of the length at rest. On the other hand, something traveling at nine-tenths the speed of light is shortened to

$$\frac{L}{L_0} = \sqrt{1 - \frac{(0.9c)^2}{c^2}} = 0.436 = 43.6\%$$

of its length at rest, a significant change.

Although the Lorentz-Fitzgerald contraction is verified by experimental measurements, the visual appearance of an object in rapid motion (if we could actually see it) would not exhibit such a shortening. This apparent contradiction comes about because light reaching the eye or a camera from the more distant parts of the object left it at an earlier time than that coming from the closer parts. What we see is therefore a composite image, since the object was at different places when the various elements that make up the image left it. This effect lengthens the apparent size of a moving object in its direction of motion, and turns out to exactly cancel the Lorentz-Fitzgerald contraction! A three-dimensional body, for instance a cube, would be seen as though rotated in orientation without any change in shape; in the limit of v approaching c the angle through which it would be turned would be 180°. A sphere would suffer no distortion whatever. We must distinguish this visual effect from the Lorentz-Fitzgerald contraction itself, which is a physical phenomenon. In the absence of this contraction, moving bodies would appear as though lengthened by a corresponding amount. Interestingly enough, this analysis of the visual appearance of rapidly moving objects was not made until 1959, which was over a half century after Einstein published the special theory of relativity.

2–6 TIME DILATION

Time intervals, too, are affected by relative motion. A clock moving with respect to an observer appears to tick less rapidly than it does when it is at rest with respect to him; that is, if we observe the length of time some event requires in a frame of reference in motion relative to us, our clock would indicate a longer time interval than a clock in the moving frame.

Let us derive this *time dilation* directly from the Lorentz transformation. Suppose a rider in the car of Fig. 2–8 is located at the point x' in his frame of reference, where, as before, the origins of the stationary frame S (the ground) and the moving frame S' (the car) coincide at $t = t' = 0$. The rider glances at his clock and finds that the time is t_1'. At the same moment a man beside the road finds the time to be t_1, where, according to Eq. (2–27),

$$t_1 = \frac{t_1' + vx/c^2}{\sqrt{1 - v^2/c^2}}.$$

Later the rider in the car again glances at his clock and notes that the time is now t_2', so that the time that has elapsed in his frame of reference is

$$t_0 = t_2' - t_1'.$$

According to the clock on the ground this time interval ended at t_2, where

$$t_2 = \frac{t_2' + vx'/c^2}{\sqrt{1 - v^2/c^2}}.$$

Hence to the man on the ground the period of time involved is

$$t = t_2 - t_1 = \frac{t_2' - t_1'}{\sqrt{1 - v^2/c^2}},$$

or

$$t = \frac{t_0}{\sqrt{1 - v^2/c^2}}. \tag{2–29}$$

An event occurring in the moving frame (the car) that occupies a time interval of t_0 is longer by the factor $1/\sqrt{1 - v^2/c^2}$ to an observer in the stationary frame (the ground).

A good illustration of both the time dilation of Eq. (2–29) and the length contraction of Eq. (2–28) occurs in the decay of unstable particles called mu-mesons, whose properties we shall discuss in greater detail in Chapter 10. For the moment what concerns us is the fact that a mu-meson decays into an electron an average of 2×10^{-6} sec after it comes into being. Now mu-mesons are created in the atmosphere thousands of meters above the ground by fast cosmic-ray particles incident upon the earth from space, and reach sea level in profusion. Their speeds are observed to be about 2.994×10^8 m/sec, or 0.998 of the velocity of light c. But in $t_0 = 2 \times 10^{-6}$ sec, the meson lifetime, they can travel a distance of only

$$h = vt_0 = 2.994 \times 10^8 \, \frac{\text{m}}{\text{sec}} \times 2 \times 10^{-6} \, \text{sec} = 599 \text{ m},$$

while they are actually created at elevations 10 or more times greater than this.

We can resolve the meson paradox by using the results of the special theory of relativity. Let us first do this from the frame of reference of the meson, where its lifetime is 2×10^{-6} sec. While the meson lifetime is unaffected by the motion, its distance to the ground appears shortened by the factor

$$\frac{h}{h_0} = \sqrt{1 - (v^2/c^2)}.$$

That is, while we, on the ground, measure the altitude of the meson as h_0, the meson sees it as h. If we let h be 599 meters, the maximum distance the meson can go in its own frame of reference at the speed $0.998c$ before decaying, we find that the corresponding distance h_0 in *our* reference frame is

$$h_0 = \frac{h}{\sqrt{1 - (v^2/c^2)}} = \frac{599 \text{ m}}{\sqrt{1 - [(0.998c)^2/c^2]}}$$

$$= \frac{599 \text{ m}}{\sqrt{1 - 0.996}} = \frac{599 \text{ m}}{0.063} = 9510 \text{ m}.$$

Hence, despite its brief life span, the meson is able to reach the ground from quite respectable altitudes.

Now let us apply the special theory of relativity from the point of view of an observer on the ground. On the ground we find the meson altitude to be h_0, but its lifetime is no longer $t_0 = 2 \times 10^{-6}$ sec; its apparent lifetime t to *us* has been extended owing to the relative motion to the value

$$t = \frac{t_0}{\sqrt{1 - (v^2/c^2)}} = \frac{2 \times 10^{-6} \text{ sec}}{\sqrt{1 - [(0.998c)^2/c^2]}}$$

$$= \frac{2 \times 10^{-6} \text{ sec}}{0.063} = 31.7 \times 10^{-6} \text{ sec},$$

almost 16 times longer than when it is at rest with respect to us. In

31.7×10^{-6} sec a meson whose speed is $0.998c$ can travel a distance

$$h_0 = vt = 2.994 \times 10^8 \, \frac{\text{m}}{\text{sec}} \times 31.7 \times 10^{-6} \, \text{sec}$$
$$= 9510 \text{ m},$$

the same distance we found previously. The two points of view give consistent results.

2-7 THE RELATIVITY OF MASS

According to relativity, measurements of the same object or event will be different for observers in relative motion. However, despite these differences, the laws of motion arrived at by the various observers must have exactly the same mathematical form. If we consider in detail a collision between two bodies as witnessed from the reference frames S and S' and impose the requirement that linear momentum be conserved, we would find that mass is no more independent of motion than space and time are. As seen from the earth, a rocket ship in flight is shorter than its twin still on the ground, its clocks tick more slowly, and, in addition, its mass is greater. If the *rest mass* of an object (that is, its mass as measured by somebody stationary with respect to it) is m_0, the mass m that will be measured by somebody moving with the speed v with respect to it is

$$m = \frac{m_0}{\sqrt{1 - (v^2/c^2)}}. \tag{2-30}$$

Since the denominator of this equation is always less than one, an object will always appear more massive when in relative motion than when at rest. This mass increase is reciprocal; to a measuring device on the rocket ship in flight a twin ship on the ground also appears to have a mass m greater than its own mass m_0.

Relativistic mass increases are significant only at speeds approaching that of light. At a speed one-tenth that of light, the mass increase amounts to only 0.5%, but this increase is over 100% at a speed of nine-tenths that of light. Only atomic particles such as electrons, protons, mesons, and so on have sufficiently high speeds for relativistic effects to take place, and in dealing with such particles, as we have seen, these effects must be carefully taken into account. Historically, the first confirmation of Eq. (2-30) was the discovery of Bucherer in 1908 that the ratio e/m of the electron's charge to its mass was smaller for fast electrons than for slow ones; this equation, like the others of special relativity, has been verified so many times that it is now accepted as one of the basic formulas of physics.

Equation (2–30) has something interesting to say about the greatest speed an object can have. The closer v approaches c, the closer v^2/c^2 approaches one, and the closer $\sqrt{1 - (v^2/c^2)}$ approaches zero. As the denominator of Eq. (2–30) becomes smaller, the mass m becomes larger, so that if the relative speed v actually were equal to the speed of light, the object's mass would be infinite. The concept of an infinite mass anywhere in the universe is, of course, nonsense on many counts; it would have required an infinite force to have accelerated it to the speed of light, its length in the direction of motion would be zero by Eq. (2–28) so that its volume would be zero, and it would exert an infinite gravitational force on all other bodies in the universe. Hence we interpret Eq. (20–30) to mean that no material body can ever equal or exceed the speed of light.

We should keep in mind that c in the formulas of relativity refers to the speed of light in free space, 3×10^8 m/sec. In all material media, such as water, glass, and air, light travels more slowly than this, and atomic particles are capable of moving with higher speeds in such media than the speed of light *in them,* though never faster than the speed of light in free space. When an electrically charged particle moves through a substance at a speed exceeding that of light in the substance, a cone of light waves is emitted in a process roughly similar to that in which a ship produces a bow wave as it moves through the water faster than water waves do. The emission of light waves under these circumstances constitutes the *Cerenkov effect.*

2–8 MASS AND ENERGY

According to special relativity, Newton's second law of motion must be written

$$\mathbf{F} = \frac{d}{dt}(m\mathbf{v}). \qquad (2\text{–}31)$$

This is not the same as $\mathbf{F} = m\mathbf{a}$ since

$$\frac{d}{dt}(m\mathbf{v}) = m\frac{d\mathbf{v}}{dt} + \mathbf{v}\frac{dm}{dt} = m\mathbf{a} + \mathbf{v}\frac{dm}{dt},$$

and the time variation of mass dm/dt is not zero if v varies with time. Hence the correct statement of the second law of motion is that the time rate of change of the momentum of an object is equal to the net force acting on it.

In relativistic mechanics, as in ordinary mechanics, the kinetic energy of a moving body is the work needed to bring it from rest to whatever speed v it happens to have. That is,

$$KE = \int_0^s F\,ds,$$

where F is the component of the applied force in the direction of the displacement ds, and s is the distance through which the force acts. Making use of Eq. (2–31) and the fact that $ds/dt = v$, we obtain

$$\text{KE} = \int_0^s \frac{d}{dt}(mv)\, ds = \int_0^{mv} \frac{ds}{dt}\, d(mv)$$

$$= \int_0^{mv} v\, d(mv) = \int_0^v v\, d\, \frac{m_0 v}{\sqrt{1 - v^2/c^2}}.$$

According to the procedure for integrating by parts,

$$\int x\, dy = xy - \int y\, dx,$$

and so, applying this formula here, we have

$$\text{KE} = \frac{m_0 v^2}{\sqrt{1 - v^2/c^2}} - m_0 \int_0^v \frac{v\, dv}{\sqrt{1 - v^2/c^2}}$$

$$= \frac{m_0 v^2}{\sqrt{1 - v^2/c^2}} + m_0 c^2 [\sqrt{1 - v^2/c^2}]_0^v$$

$$= \frac{m_0 c^2}{\sqrt{1 - v^2/c^2}} - m_0 c^2 = (m - m_0)c^2. \tag{2–32}$$

Equation (2–32) asserts that the kinetic energy of an object is equal to the increase in its mass $m - m_0$ due to its motion multiplied by the square of the speed of light c.

If we rewrite Eq. (2–32) as

$$mc^2 = \text{KE} + m_0 c^2, \tag{2–33}$$

we see that we can interpret mc^2 as the total energy of a body in motion relative to an observer who measures the body's mass as m, provided that the quantity $m_0 c^2$ is the energy the body has when it is at rest relative to this observer. This interpretation has turned out to be correct, and the energy

$$E_0 = m_0 c^2 \tag{2–34}$$

is called the *rest energy* of the body. Thus a 1-kg mass is equivalent to 9×10^{16} joules. Even a minute bit of matter can yield an immense amount of energy. In fact, the conversion of matter into energy is the energy source of the sun and stars as well as that of atomic and hydrogen bombs of chemical reactions, and, indeed, of *all* processes in which energy is liberated. The equivalence of mass and energy, a direct consequence of the postulates of special relativity, is the most famous of the various results Einstein obtained early in this century.

About 4 billion kg of matter is converted into energy in the sun per second. The energy equivalent of 4 billion kg of matter is

$$E_0 = m_0 c^2 = 4 \times 10^9 \text{ kg} \times (3 \times 10^8 \text{ m/sec})^2$$
$$= 3.6 \times 10^{26} \text{ j}.$$

Since 1 j/sec = 1 watt and 3.6×10^{26} j are evolved by the sun in each second, its power output is

$$P = 3.6 \times 10^{26} \text{ watts}.$$

When the relative speed v is small compared with c, we expect that the formula

$$\text{KE} = \frac{m_0 c^2}{\sqrt{1 - v^2/c^2}} - m_0 c^2$$

for kinetic energy should become just the $\frac{1}{2} m_0 v^2$ we are familiar with and which has been verified by experiment at low speeds. Let us see whether this is true. In general, the binomial theorem of algebra tells us that if some quantity x is much smaller than 1,

$$(1 + x)^n = 1 + nx.$$

Here, with

$$x = \frac{-v^2}{c^2} \quad \text{and} \quad n = -\tfrac{1}{2},$$

we find that the above formula indeed becomes

$$\text{KE} = \left(1 + \frac{\tfrac{1}{2} v^2}{c^2}\right) m_0 c^2 - m_0 c^2 = \tfrac{1}{2} m_0 v^2.$$

In the foregoing calculation, relativity has met an important test; it has yielded exactly the same results as those of ordinary mechanics at low speeds, where we know by experience that the latter are perfectly valid. It is proper, then, to think of the principles of mechanics as no more than approximations to the correct relativistic principles of mechanics, valid approximations in nearly all common situations but requiring modification under certain circumstances.

2–9 VELOCITY ADDITION

We have already mentioned that the Galilean velocity transformation of Eqs. (2–9) through (2–11) is not consistent with the constancy of the speed of light, since it predicts that different observers will find different values of c depending upon their states of motion. To obtain the correct transformation all that is necessary is to differentiate the appropriate Lorentz transformation equations. For instance, we might wish to relate v_x', v_y', v_z', the components of the velocity of an object as measured in the moving S'-system, to the components v_x, v_y, v_z that would be measured in the stationary S-system. From Eqs. (2–24) through (2–27), noting that both the relative velocity v (assumed in the x-direction) and the speed of light c are constant, we have

$$dx = \frac{dx' + v\, dt'}{\sqrt{1 - v^2/c^2}}, \qquad dy = dy',$$

$$dz = dz', \qquad dt = \frac{dt' + v\, dx'/c^2}{\sqrt{1 - v^2/c^2}}.$$

Hence

$$v_x = \frac{dx}{dt} = \frac{dx' + v\, dt'}{dt' + v\, dx'/c^2},$$

and dividing through by dt' yields

$$v_x = \frac{dx'/dt' + v}{1 + v/c^2\, dx'/dt'} = \frac{v_x' + v}{1 + vv_x'/c^2}. \qquad (2\text{–}35)$$

In a similar manner we obtain the other two components of velocity,

$$v_y = \frac{v_y'\sqrt{1 - v^2/c^2}}{1 + vv_x'/c^2} \qquad (2\text{–}36)$$

and

$$v_z = \frac{v_z'\sqrt{1 - v^2/c^2}}{1 + vv_x'/c^2}. \qquad (2\text{–}37)$$

When v is small compared with c, these equations reduce to the Galilean transformation of Eqs. (2–9) through (2–11).

Let us apply the above results to a pulse of light emitted from the moving system S'. If the light is sent out in the $+x$-direction, $v_x' = c$, and

$$v_x = \frac{c + v}{1 + vc/c^2} = \frac{c(c + v)}{c + v} = c;$$

the measured speed of light is exactly the same in both systems. This conclusion further verifies the consistency of the Lorentz transformation with the postulates of relativity. Ordinary mechanics, of course, predicts that the speed of the pulse measured in S would be $c + v$.

Since nothing can move faster than the speed of light, it might seem impossible to pass a space ship whose speed with respect to the earth is $0.9c$ at a relative speed of, say, $0.5c$. Ordinary mechanics indicates that the required speed is $0.9c + 0.5c = 1.4c$. If we set $v_x' = 0.5c$ and $v = 0.9c$ in Eq. (2–35), however, we obtain

$$v_x = \frac{1.4c}{1 + 0.45} = 0.966c. \cdot$$

It is necessary to move less than 8 percent faster than a space ship whose speed is $0.9c$ in order to pass it at a relative speed of half the speed of light.

PROBLEMS

1. Can an observer in a windowless laboratory in principle determine whether the earth is (a) moving through space with a uniform velocity, (b) moving through space with a uniform linear acceleration, (c) rotating on its axis?

2. The hypothetical speed of the earth through the ether is its orbital speed of 3×10^4 m/sec. (a) If light takes precisely 3×10^{-7} sec to travel through the Michelson-Morley apparatus in the direction parallel to this motion, how long will it take to travel through it in the direction perpendicular to this motion? (b) If light of wavelength 6×10^{-7} m is used, what fraction of a wavelength does the difference in the above times represent?

3. Two observers, A on earth and B in a rocket ship whose speed is 2×10^8 m/sec, both set their watches to 1:00 when the ship is abreast of the earth. (a) When A's watch reads 1:30 he looks at B's watch through a telescope. What does it read? (b) When B's watch reads 1:30 he looks at A's watch through a telescope. What does it read?

4. A certain process requires 10^{-6} sec to occur in an atom at rest in the laboratory. How much time will this process require to an observer in the laboratory when the atom is moving at a speed of 5×10^7 m/sec?

5. How fast would a rocket ship have to go for each year on the ship to correspond to two years on the earth?

6. What is the relativistic contraction of an airplane 100 ft long when it moves directly away from an observer at 400 mi/hr?

7. How fast would a rocket ship have to go relative to an observer for its length to be contracted to 99% of its length when at rest?

8. The specific heat of lead is 0.03 kcal/kg·°C. If a bar of lead has a mass of 100 kg at 20°C, how much mass will it gain when heated to 300°C?

9. How much mass is lost by 1 kg of water at 0°C when it turns to ice at 0°C?

10. Compute the rest energy of a proton in joules and in electron volts.

11. Dynamite liberates 1.3×10^3 kcal/kg when it explodes. What percentage of its total energy content is this?

12. Typical chemical reactions absorb or release energy at the rate of several electron volts per molecular change. What change in mass is associated with the absorption of 1 ev of energy? With the release of 1 ev of energy?

13. What speed must a body have if its mass is to double?

14. (a) How much energy is required to double the speed of an electron whose initial speed is 7×10^7 m/sec? (b) How much energy is required to double the speed of an electron whose initial speed is 1.4×10^8 m/sec?

15. The relativistic increase in mass of moving protons limits the ability of a cyclotron to accelerate them. (a) At what speed will a proton's mass exceed its rest mass by 1%? (b) What kinetic energy in electron volts does this speed correspond to?

16. An electron is accelerated through a potential difference of 10^5 volts. (a) What is its speed? (b) What is its mass? (c) Compare the answers to (a) and (b) with the values that classical physics would predict.

17. The electron beam in a television picture tube can move across the screen faster than the speed of light. Why does this not violate the special theory of relativity?

18. How many joules of energy per kilogram of rest mass are required to accelerate a space ship to a speed of 2.94×10^8 m/sec?

19. Show that the second law of motion can be expressed in terms of the rest mass and speed of a particle being acted upon by a force as

$$\mathbf{F} = m_0(d\mathbf{v}/dt)/(1 - v^2/c^2)^{3/2}.$$

20. A lady on Venus sees two space ships approaching her from opposite directions at speeds of $0.8c$ and $0.9c$ respectively. What is the relative speed of the two space ships as measured by an observer on either of them?

3

Quanta

"I therefore tried immediately to weld the elementary quantum of action, h, somehow into the framework of classical theory. But in the face of all such attempts the constant showed itself to be obdurate. So long as it could be regarded as infinitesimally small, i.e., when dealing with higher energies and longer periods of time, everything was in perfect order. But in the general case difficulties would arise at one point or another, difficulties which became more noticeable as higher frequencies were taken into consideration. The failure of every attempt to bridge that obstacle soon made it evident that the elementary quantum of action plays a fundamental part in atomic physics and that its introduction opened up a new era in natural science, for it heralded the advent of something entirely unprecedented and was destined to remodel basically the physical outlook and thinking of man which, ever since Leibniz and Newton laid the ground work for infinitesimal calculus, were founded on the assumption that all causal interactions are continuous."

Max Planck (1858–1947)

The formulation of the theory of relativity and that of the quantum theory of light, both of which took place early in this century, profoundly altered our ways of thinking about the physical world. In the previous chapter we examined some of the remarkable consequences of relativity, consequences so alien to our intuition that our first feeling must be one of astonishment. Now we come to the realm of quanta, which will be no less remarkable or astonishing. But these subjective terms are really not justified, because they are merely based upon the limits to our imagination that are imposed by our experience; if we were in a world where we were about the same size as an electron, relativistic and quantum phenomena would be familiar (though then most macroscopic phenomena would not). We shall find that the ideas presented in this chapter are not only intrinsically of great interest, but are also essential if we are to understand the material on the atom and the nucleus that follows.

3–1 THE PHOTOELECTRIC EFFECT

Toward the end of the 19th century a number of experiments were performed that revealed the emission of electrons from a metal surface when light (particularly ultraviolet light) is shined on it (Fig. 3–1). This

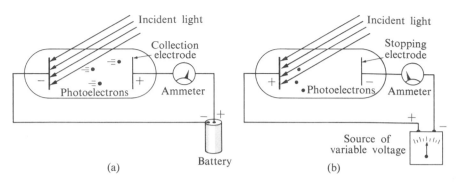

FIG. 3–1. (a) A method of detecting the photoelectric effect. The photoelectrons ejected from the metal plate being irradiated are attracted to the positive collection electrode at the other end of the tube, and the current that results is measured with the ammeter. (b) A method of detecting the maximum energy of the photoelectrons. As the stopping electrode is made more negative, the slower photoelectrons are repelled before they can reach it. Finally a voltage will be reached at which no photoelectrons whatever are received at the stopping electrode, as indicated by the current dropping to zero, and this voltage corresponds to the maximum photoelectron energy.

phenomenon is known as the *photoelectric effect*. It is not, at first glance, anything to surprise us, for light waves carry energy, and some of the energy absorbed by the metal may somehow concentrate on individual electrons and reappear as kinetic energy. Upon closer inspection of the data, however, we find that the photoelectric effect can hardly be explained in so straightforward a manner.

The first peculiarity of the photoelectric effect is that the energy of the emitted electrons (which are called *photoelectrons*) does not depend upon the intensity of the light. A bright light yields more electrons than a dim one, but their average energy remains the same. Furthermore, even when the metal surface is faintly illuminated, the photoelectrons leave the surface immediately. This behavior contradicts the electromagnetic theory of light, which predicts that the energy of photoelectrons should depend upon the intensity of the light beam responsible for them and that, if the beam has a very low intensity, a certain period of time (perhaps even days) must elapse on the average before any individual electrons accumulate enough energy to leave the metal.

Still more puzzling is the fact that the photoelectron energy depends upon the *frequency* of the light employed. At frequencies below a certain critical one (which is characteristic of the particular metal), no electrons whatever are emitted. Above this threshold frequency, the photoelectrons have a range of energies from zero to a certain maximum value, and *this maximum energy increases with increasing frequency*. High frequencies result in high maximum photoelectron energies, low frequencies in low maximum photoelectron energies. Thus a faint blue light produces elec-

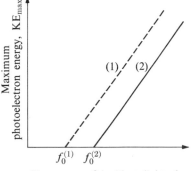

FIG. 3–2. The variation of maximum photoelectron energy with the frequency of the incident light for two target metals. No photoelectrons are emitted for frequencies less than $f_0^{(1)}$ in the case of metal (1) and less than $f_0^{(2)}$ in the case of metal (2). In both cases, however, the angle between the experimental line and either axis is the same. Hence we may write the equation of the lines as $KE_{max} = h(f - f_0)$, where h has the same value in all cases but f_0, the minimum frequency required for photoelectric emission to occur, depends upon the nature of the target metal.

trons with more energy than those produced by a bright red light, although the latter yields a greater number of them.

Figure 3–2 is a graph of maximum photoelectron energy plotted against the frequency f of the incident light in a particular experiment. It is clear that the relationship between maximum photoelectron energy, KE_{max}, and light frequency is a simple one, which we can express in equation form as

$$KE_{max} = h(f - f_0) = hf - hf_0, \tag{3–1}$$

where f_0 is the threshold frequency below which no photoemission occurs and h is a constant. Significantly, the value of h is the same in all cases, although f_0 varies with the particular metal being illuminated.

3–2 PLANCK'S RADIATION LAW

Aware that the electromagnetic theory of light, despite its notable success in accounting for other optical phenomena, failed to explain the photoelectric effect, Albert Einstein in 1905 sought some other way of understanding it. He found what he wanted in a strange assumption that Max Planck, a German physicist, had had to make five years earlier in order to explain the experimental data on blackbody radiation. As we learned in Chapter 1, classical physics could go no further than the Rayleigh-Jeans formula for the spectral energy density of this radiation, a formula whose obvious incorrectness—the "ultraviolet catastrophe"—was most disturbing. Planck introduced the idea that electromagnetic

radiation of a particular frequency f is emitted and absorbed in finite
bursts of energy

$$E = hf, \tag{3-2}$$

where h is a universal constant, and not continuously as required by the
electromagnetic theory of Maxwell. These bursts of energy are called
quanta. The value of h, which is known today as *Planck's constant*, is

$$h = 6.63 \times 10^{-34} \text{ j·sec},$$

and so each quantum contains only a minute amount of energy.

From the assumption that radiant energy is quantized, Planck was
able to show that the average energy $\overline{E}$ per mode of vibration in the
cavity equivalent of a blackbody is

$$\overline{E} = \frac{hf}{e^{hf/kT} - 1}. \tag{3-3}$$

Thus $\overline{E}$ varies with frequency instead of having the invariable classical
value of $\overline{E} = kT$. From Chapter 1 we recall that there are

$$dn_\lambda = 8\pi \frac{d\lambda}{\lambda^4}$$

modes of vibration per unit volume with wavelengths between λ and
$\lambda + d\lambda$, which is equal to

$$dn_f = \frac{8\pi}{c^3} f^2 \, df$$

modes with frequencies between f and $f + df$. Using Planck's formula
for $\overline{E}$, we find that the total energy per unit volume in this frequency
interval is

$$E_f \, df = \frac{8\pi h}{c^3} \frac{f^3 \, df}{e^{hf/kT} - 1}. \tag{3-4}$$

Equation (3–3), the *Planck radiation law*, is in perfect agreement with
all measurements of blackbody radiation. In terms of wavelength,

$$E_\lambda \, d\lambda = \frac{8\pi hc}{\lambda^5} \frac{d\lambda}{e^{hc/\lambda kT} - 1}. \tag{3-5}$$

There is no ultraviolet catastrophe any longer, since $E_\lambda \, d\lambda$ does not approach infinity as λ approaches zero, and both the Stefan-Boltzmann
formula and the Wien displacement law can be derived directly from
Eq. (3–4).

At low frequencies (long wavelengths), when hf is much smaller than
kT, Planck's expression for $\overline{E}$ reduces to the classical value of kT, and

Planck's radiation law reduces to the Rayleigh-Jeans formula. To verify this statement we note that

$$e^x = 1 + x + \frac{x^2}{2!} + \frac{x^3}{3!} + \cdots,$$

so that, when x is small, the exponential may be replaced by

$$e^x = 1 + x.$$

Hence, when $hf \ll kT$, we have

$$\overline{E} = \frac{hf}{e^{hf/kT} - 1} = \frac{hf}{(1 + hf/kT) - 1} = kT.$$

Just as the results of relativity reduce to those of classical physics when v is small compared with c, so the results of quantum theory reduce to those of classical physics when hf is small compared with kT.

3–3 QUANTUM THEORY OF LIGHT

While the energy radiated by a heated object must be assumed to come off intermittently, in order for theory and experiment to agree, Planck held to the conventional view that it nevertheless propagates through space continuously. Einstein saw that Planck's idea could be used to interpret the photoelectric effect if light not only is emitted a quantum at a time but also travels as quanta. Then the h of Eq. (3–1) is the same as the h of (3–2), and the significance of the former equation becomes clear when it is rewritten

$$hf = \text{KE}_{\max} + hf_0. \tag{3–6}$$

What this equation states is simply that

Quantum energy = maximum electron energy
+ energy required to eject an electron.

The reason for a threshold frequency is clear: it corresponds to the energy required to dislodge an electron from the metal surface. (There must be such a minimum energy, or electrons would leave metals all the time.) And, of course, there are several plausible reasons why not all photoelectrons have the same maximum energy. For instance, not all of the quantum energy hf may be transferred to a single electron, and an electron may lose some of its initial energy in collisions within the metal before it actually emerges from the surface.

Let us apply Eq. (3–6) to a specific problem. The threshold frequency for copper is $1.1 \times 10^{15} \text{ sec}^{-1}$. When ultraviolet light of frequency $1.5 \times 10^{15} \text{ sec}^{-1}$ is shined on a copper surface, what is the maximum energy of the photoelectrons? We find that

$$\text{KE}_{\text{max}} = hf - hf_0 = h(f - f_0)$$
$$= 6.63 \times 10^{-34} \, \text{j·sec} \times (1.5 - 1.1) \times 10^{15} \, \text{sec}^{-1}$$
$$= 2.7 \times 10^{-19} \, \text{j}.$$

Since 1 ev $= 1.6 \times 10^{-19}$ joule, the maximum photoelectron energy is

$$\frac{2.7 \times 10^{-19} \, \text{j}}{1.6 \times 10^{-19} \, \text{j/coul}} = 1.7 \, \text{ev}.$$

Einstein's notion that light travels as a series of little packets of energy (sometimes referred to as quanta, sometimes as *photons*) is in complete contradiction with the wave theory of light (Fig. 3–3). And the wave theory, as we know, has some powerful observational evidence on its side. There is no other way of explaining interference and diffraction effects, for example. According to the wave theory, light spreads out from a source in a manner analogous to the spreading out of ripples on the surface of a lake when a stone is dropped into it, with the energy of the light distributed continuously throughout the wave pattern. According to the quantum theory, light spreads out from a source as a succession of localized packets of energy, each sufficiently small to permit its being absorbed by a single electron. Yet, despite the particle picture of light that it presents, the quantum theory requires a knowledge of the light frequency f, a wave quantity, in order to determine the energy of each quantum.

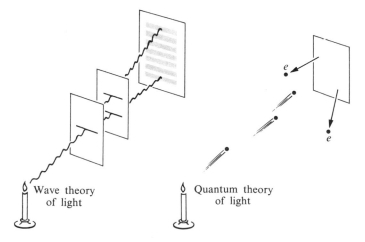

Wave theory
of light

Quantum theory
of light

FIG. 3–3.   The wave theory of light is necessary to explain diffraction and interference phenomena, which the quantum theory cannot explain. The quantum theory of light is necessary to explain the photoelectric effect, which the wave theory cannot explain.

On the other hand, the quantum theory of light is able to explain the photoelectric effect. It predicts that the maximum photoelectron energy should depend upon the frequency of the incident light and not upon its intensity, precisely the opposite of what the wave theory suggests, and it is able to explain why even the feeblest light can lead to the immediate emission of photoelectrons. The wave theory can give no reason why there should be a threshold frequency below which no photoelectrons are observed, no matter how strong the light beam, something that follows naturally from the quantum theory.

Which theory is correct? The history of physics is filled with examples of physical ideas that required revision or even replacement when new empirical data conflicted with them, but this is the first occasion in which two completely different theories are both required to explain a single physical phenomenon. In thinking about this, it is important for us to note that, in a particular situation, light behaves *either* as though it has a wave nature *or* a particle nature. While light sometimes assumes one guise and sometimes the other, there is no physical process in which both are exhibited. The same light beam can diffract around an obstacle and then impinge on a metal surface to eject photoelectrons, but these two processes occur separately. The electromagnetic theory of light and the quantum theory of light complement each other; each by itself is "correct" in certain experiments, and there are no relevant experiments which neither can explain. Light must be thought of as a phenomenon that sometimes manifests itself as particles and sometimes as waves; while we cannot visualize its "true nature," these complementary theories of light are able to account for its behavior, and we have no choice but to accept them both.

3–4 X-RAYS

If photons of light can give up their energy to electrons, can the kinetic energy of moving electrons be converted into photons? The answer is that such a transformation is not only possible, but had in fact been discovered (though not understood) prior to the work of Planck and Einstein. In 1895 Roentgen found that a mysterious, highly penetrating radiation is emitted when high-speed electrons impinge on matter. The x-rays (so called because their nature was then unknown) caused phosphorescent substances to glow, exposed photographic plates, traveled in straight lines, and were not affected by electric or magnetic fields. The more energetic the electrons, the more penetrating the x-rays, and the greater the number of electrons, the greater the density of the resulting x-ray beam.

After over 10 years of study, it was finally established that x-rays exhibit, under certain circumstances, both interference and polarization

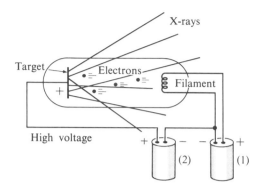

FIG. 3–4. Diagram of an x-ray tube.

effects, leading to the conclusion that they are electromagnetic waves. From the interference experiments their frequencies were found to be very high, above those in ultraviolet light.

Figure 3–4 is a diagram of an x-ray tube. Battery (1) sends a current through the filament, heating it until it emits electrons. These electrons are then accelerated toward a metallic target by the potential difference V provided by battery (2). The tube is evacuated to permit the electrons to reach the target unimpeded. The impact of the electrons causes the evolution of x-rays from the target.

What is the physical process involved in the production of x-rays? It is known that charged particles emit electromagnetic waves whenever they are accelerated, and so we may reasonably identify x-rays as the radiation accompanying the slowing down of fast electrons when they strike a metal. The great majority of the incident electrons, to be sure, lose their kinetic energy too gradually for x-rays to be evolved, and merely act to heat the target. (Consequently the targets in x-ray tubes are made of metals with high melting points, and a means for cooling the target is often provided.) A few electrons, however, lose much or all of their energy in single collisions with target atoms, and this is the energy that appears as x-rays. In other words, we may regard x-ray production as an inverse photoelectric effect. The highest frequency f_{max} found in the x-rays emitted from a particular tube should therefore correspond to a quantum energy of hf_{max}, where hf_{max} equals the kinetic energy of the electrons. That is,

$$\text{KE} = hf_{max}.$$

If we denote the charge of an electron by e, its kinetic energy after being accelerated through a potential difference V is

$$\text{KE} = eV;$$

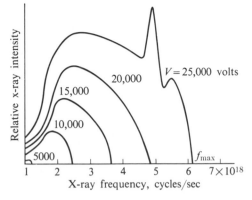

FIG. 3–5. X-ray spectra of molybdenum for accelerating potentials from 5000 to 25,000 v.

hence we expect that

$$hf_{max} = eV \tag{3–7}$$

in the operation of an x-ray tube.

A conventional x-ray machine might have an operating potential of 50,000 volts. To find the highest frequency present in its radiation, we use Eq. (3–7):

$$f_{max} = \frac{eV}{h} = \frac{1.6 \times 10^{-19} \text{ coul} \times 5 \times 10^4 \text{ volts}}{6.6 \times 10^{-34} \text{ j·sec}}$$

$$= 1.2 \times 10^{19} \text{ sec}^{-1}.$$

Figure 3–5 shows the x-ray spectra that result when a molybdenum target is bombarded by electrons at several different accelerating voltages. The proportionality between f_{max} and V predicted by Eq. (3–7) agrees with the measurements, which provides confirmation for the quantum theory of light. The curves show that all frequencies below f_{max} are emitted, as we expect, and that their intensities vary in a regular manner. In addition, the highest-voltage curve possesses curious, sharp "spikes" that indicate the enhanced production of certain frequencies. These spikes originate in rearrangements of the electron structures of the target atoms that are induced by the bombarding electrons. A number of specific energies may be involved in these rearrangements, depending upon the particular target material, so that the presence of discrete x-ray frequencies has a straightforward explanation.

3–5 THE COMPTON EFFECT

In the photoelectric effect, a light photon gives up most or all of its energy to an electron, leading us to attribute particle properties to the photon. How far can we carry the analogy? For instance, can we analyze

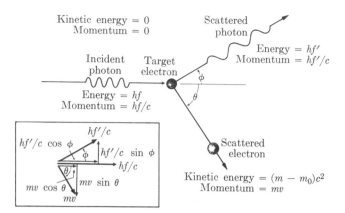

FIG. 3–6. The Compton effect. A vector diagram showing the various momenta and their projections parallel and perpendicular to the original photon direction is included to help in understanding the origin of Eqs. (3–10) and (3–11).

the case of a photon making a glancing collision with an electron just as if both were billiard balls? Let us see what we can expect in such a collision if, indeed, photons behave like particles. Figure 3–6 shows the collision; an x-ray photon strikes an electron, which is initially at rest, and is deflected to one side, while the electron starts to move. As a result of the impact, the photon loses an amount of energy equal to the kinetic energy $(m - m_0)c^2$ gained by the electron. (The relativistic formula for kinetic energy is the proper one to use here.) If the original frequency associated with the photon is f, its loss of energy means that its subsequent frequency will be the lower value f', where

$$\text{Change in photon energy} = \text{electron kinetic energy}$$

$$hf - hf' = (m - m_0)c^2. \tag{3–8}$$

A moving electron has momentum mv as well as energy, so a photon must have momentum if it is to transfer any to an electron. The momentum of a photon, if it behaves like a particle, should be

$$\text{Photon momentum} = m_p c$$

since it travels with the speed of light c. While a photon has no rest mass, we can take its energy hf as equal to the energy $m_p c^2$ of a particle whose relativistic mass is m_p. Hence

$$hf = m_p c^2, \qquad m_p = \frac{hf}{c^2},$$

and the photon momentum is

$$m_p c = \frac{hf}{c}. \tag{3–9}$$

Momentum is a vector quantity, possessing direction as well as magnitude, and we must separately conserve momentum in the original direction of the photon and in a direction perpendicular to it in the plane of the electron and scattered photon. The initial momentum of the photon is hf/c, its final momentum is hf'/c, and the initial and final momenta of the electron are respectively zero and mv. In the original photon direction, as Fig. 3–6 indicates,

$$\text{Initial momentum} = \text{final momentum}$$

$$\frac{hf}{c} + 0 = \frac{hf'}{c} \cos \phi + mv \cos \theta, \tag{3–10}$$

and perpendicular to this direction,

$$\text{Initial momentum} = \text{final momentum}$$

$$0 = \frac{hf'}{c} \sin \phi - mv \sin \theta. \tag{3–11}$$

When Eqs. (3–8), (3–10), and (3–11) are solved, the new photon frequency f' is found to be related to the original frequency f and to the angle ϕ between the initial photon direction and the scattered photon direction by the formula

$$\frac{1}{f'} = \frac{1}{f} + \frac{h}{m_0 c^2} (1 - \cos \phi), \tag{3–12}$$

where m_0 is the rest mass of the electron. This formula can readily be checked by experiment; a beam of x-rays of a single, known frequency f is directed at some target, and the frequencies of the scattered x-rays measured at various angles. According to Eq. (3–12) the scattered x-rays should have lower frequencies than those in the original beam, with the change in frequency greatest for the greatest scattering angle. Experiments of this kind were performed by A. H. Compton in the early 1920's, and he was led to derive Eq. (3–12) in order to account for the results. Theory and experiment agree, and the phenomenon they refer to has become known as the *Compton effect*.

3–6 MATTER WAVES

As we have seen, electromagnetic waves under certain circumstances have properties indistinguishable from those of particles. It requires no greater stretch of the imagination to speculate whether what we normally think of as particles might not have wave properties, too. This speculation was first made in 1924 by Louis de Broglie. He started with the formula for the momentum of a photon,

$$\text{Momentum} = \frac{hf}{c},$$

which can be expressed in terms of wavelength λ as

$$\text{Momentum} = \frac{h}{\lambda}$$

since

$$\lambda f = c.$$

Hence, for a photon,

$$\lambda = \frac{h}{\text{momentum}}.$$

De Broglie suggested that this equation for wavelength is a perfectly general one, applying to material particles as well as to photons. In the case of particles,

$$\text{Momentum} = mv,$$

and so the *de Broglie wavelength* of a particle is

$$\lambda = \frac{h}{mv}. \tag{3–13}$$

The higher the momentum of the particle, the shorter its wavelength; $m_0/\sqrt{1 - (v^2/c^2)}$ must be used for m in Eq. (3–13).

How can de Broglie's hypothesis be verified? Perhaps the most striking exhibition of wave behavior is a diffraction pattern, which depends upon the ability of waves both to bend around obstacles and to interfere constructively and destructively with one another. Several years after de Broglie's work, Davisson and Germer in the United States and G. P. Thomson in England independently demonstrated that streams of electrons are diffracted when they are scattered from crystals whose atoms have appropriate spacing. The diffraction patterns they observed were in complete accord with the electron wavelength predicted by Eq. (3–13). As is true for electromagnetic waves, the wave and particle aspects of moving bodies never appear together in the same experiment, so we cannot determine which is the "correct" description. All we can say is that at certain times we must think of a moving body as a particle (in the sense of Newton's laws of motion) and at other times as a wave; whatever the ultimate nature of matter, its manifestations to us exhibit this inescapable duality.

Let us calculate the de Broglie wavelengths of two moving objects, an automobile whose speed is 60 mi/hr and an electron whose speed is 10^7 m/sec. The automobile might weigh 3200 lb, corresponding to a mass of 100 slugs, and 60 mi/hr $=$ 88 ft/sec; hence,

$$\lambda = \frac{h}{mv} = \frac{6.62 \times 10^{-34}\,\text{j·sec} \times 0.738\,\text{ft·lb/j}}{100\,\text{slugs} \times 88\,\text{ft/sec}} = 5.5 \times 10^{-38}\,\text{ft.}$$

The wavelength of the automobile is so small relative to its dimensions that we expect no wave aspects in its behavior. For the electron, however,

$$\lambda = \frac{h}{mv} = \frac{6.62 \times 10^{-34}\,\text{j·sec}}{9.1 \times 10^{-31}\,\text{kg} \times 10^{7}\,\text{m/sec}} = 7.3 \times 10^{-9}\,\text{m,}$$

which is of the order of magnitude of atomic dimensions. It is not surprising, then, that electrons with velocities of about 10^{7} m/sec exhibit diffraction effects when incident upon crystals.

3–7 THE UNCERTAINTY PRINCIPLE

The wave nature of particles and the particle nature of waves have a rather unexpected concomitant. Suppose that we wish to measure the position and momentum of something at a particular instant. To do so, we must touch it with something else that will carry the required information back to us; that is, we must poke it with a stick, shine light on it, or perform some similar act. Suppose we are looking at an electron with light of wavelength λ, as in Fig. 3–7. Each photon of this light has the momentum h/λ. When one of these photons bounces off the electron (which must occur if we are to "see" it), the electron's original momentum will be changed. The exact amount of the change Δmv cannot be predicted, but it will be of the same order of magnitude as the photon momentum h/λ. Hence

$$\Delta mv \approx h/\lambda; \qquad\qquad (3\text{--}14)$$

the *larger* the wavelength, the smaller the uncertainty in momentum. Because light is a wave phenomenon, we cannot expect to determine the

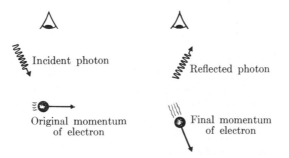

FIG. 3–7. An example of the uncertainty principle: it is impossible to accurately determine the position and momentum of a body at the same time.

electron's position with perfect accuracy even with the best of instruments; a reasonable estimate of the irreducible uncertainty Δx in the measurement might be one wavelength. That is,

$$\Delta x \approx \lambda; \qquad (3\text{--}15)$$

the *smaller* the wavelength, the smaller the uncertainty in position. Hence if we use light of short wavelength to increase the accuracy of our position measurement, there will be a corresponding decrease in the accuracy of our momentum measurement, while light of long wavelength will yield an accurate momentum but an inaccurate position.

From Eqs. (3–14) and (3–15) we see that, as a general rule,

$$\Delta x\, \Delta mv \geq h \qquad (3\text{--}16)$$

which is one form of the *uncertainty principle* first obtained by Werner Heisenberg in 1927. It states that the product of the uncertainty Δx in the position of a body at some instant and the uncertainty Δmv in its momentum at the same instant is equal to or greater than Planck's constant. In other words, we cannot ever hope to simultaneously measure both position and momentum with perfect accuracy. The uncertainty is not in our instruments but in nature. On a macroscopic scale, since h is so minute, the limitation imposed upon measurements by the uncertainty principle is negligible, but on a microscopic scale the uncertainty principle dominates many phenomena.

A significant application of the uncertainty principle is to the question of whether electrons are present within atomic nuclei. As we shall learn in Chapter 8, nuclei are roughly 10^{-15} meter across. If an electron is to be confined within a nucleus, the uncertainty in its position cannot exceed 10^{-15} meter. The consequent inherent uncertainty in the electron's momentum is

$$\Delta(mv) \geq \frac{h}{\Delta x} \geq \frac{6.62 \times 10^{-34}\, \text{j·sec}}{10^{-15}\, \text{m}}$$

$$\geq 6.62 \times 10^{-19}\, \frac{\text{kg·m}}{\text{sec}}.$$

This momentum uncertainty corresponds to an energy uncertainty of over 10^9 ev, which is a billion electron volts! But there is no evidence that electrons found in atoms ever have much more than 1/1000 of this huge amount of energy, and we therefore conclude that electrons are not present inside atomic nuclei.

Another form of the uncertainty principle is easily obtained. Suppose that we are measuring the energy E emitted in some process during the period of time Δt. If the energy is in the form of electromagnetic waves, the time Δt limits the accuracy with which we can measure the wave frequency f. Because the uncertainty in the number of waves we count in a given wave train is of the order of one wave, and

$$\text{Frequency} = \frac{\text{number of waves}}{\text{time interval}},$$

the uncertainty in frequency Δf of our determination is

$$\Delta f \approx 1/\Delta t.$$

The corresponding uncertainty in energy ΔE is

$$\Delta E \approx h \, \Delta f,$$

and so

$$\Delta E \geq \frac{h}{\Delta t}$$

or

$$\Delta E \, \Delta t \geq h. \tag{3-17}$$

Equation (3–17) states that the product of the uncertainty in a measured amount of energy and the time available for the measurement is at best equal to Planck's constant.

3–8 CAUSALITY

Until now we have taken for granted the notion of causality, namely that a given cause always produces a specific effect. The "laws" of physics that we have studied are all cause-and-effect relationships. For example, $\mathbf{F} = m\mathbf{a}$, Newton's second law of motion, states that the acceleration a particle experiences as a result of the application of a force is invariably proportional to the magnitude of the force and inversely proportional to the particle's mass. But, because of the uncertainty principle, it is impossible ever to perform experiments of infinite accuracy, and therefore impossible ever to establish that $\mathbf{F}$ is invariably equal to precisely $m\mathbf{a}$.

How seriously can we take relationships whose truth cannot possibly be verified? The answer is not as clear-cut as we would prefer. While causal relations break down when applied to individual elementary particles, and hence cannot be regarded as expressions of *truth*, they may nevertheless be regarded as expressions of *probability*. If sufficiently accurate measurements could be made of the responses of electrons to applied forces, we would expect to find that $\mathbf{F} = m\mathbf{a}$ is obeyed *on the average*, but that the individual data would show deviations from this behavior. The ordinary material bodies we encounter in our lives are aggregates of such vast numbers of particles that their average behavior is all that we perceive. Hence causality is a reasonably reliable guide to macroscopic events, but we should not expect it to hold in applications remote from our direct experience.

PROBLEMS

1. If light transfers energy by means of separate quanta, why do we not perceive a faint light as a series of tiny flashes?

2. Yellow light has a wavelength of 6×10^{-7} m. How many photons are emitted per second by a yellow lamp radiating at a power of 10 watts? (1 watt = 1 j/sec.)

3. A radio transmitter operates at a frequency of 880 kc/sec (880×10^3 cycles/sec) and a power of 10,000 watts. How many photons per second does it emit?

4. What is the lowest-frequency light that will cause the emission of photoelectrons from a surface whose nature is such that 1.9 ev are required to eject an electron?

5. Photoelectrons are emitted with a maximum speed of 7×10^5 m/sec from a surface when light of frequency 8×10^{14} cycles/sec is shined on it. What is the threshold frequency for this surface?

6. Light of wavelength 5×10^{-7} m falls on a potassium surface whose nature is such that 2 ev are needed to eject an electron. What is the maximum kinetic energy in electron volts of the photoelectrons that are emitted?

7. Why do you think the wave aspect of light was discovered earlier than its particle aspect?

8. What potential difference must be applied across an x-ray tube for it to emit x-rays with a minimum wavelength of 10^{-11} m?

9. What potential difference must be applied across an x-ray tube for it to emit x-rays with a maximum frequency of 10^{19} cycles/sec?

10. A target is bombarded by 8×10^4 ev electrons. What is the highest frequency present in the emitted x-rays?

11. Electrons are accelerated in television tubes through potential differences of about 10,000 volts. Find the highest frequency of the electromagnetic waves that are emitted when these electrons strike the screen of the tube. Use Fig. 21–5 to determine what type of waves these are.

12. Find the energy and momentum of an x-ray photon whose frequency is 5×10^{18} cycles/sec.

13. Find the energy and momentum of an x-ray photon whose wavelength is 2×10^{-11} m.

14. An x-ray photon of frequency 3×10^{19} cycles/sec is scattered through an angle of 90° after colliding with a stationary electron. (a) What is its new frequency? (b) What is the kinetic energy of the electron after the collision?

15. An x-ray photon of frequency 10^{19} cycles/sec is scattered through an angle of 45° after colliding with a stationary electron. (a) What is its new frequency? (b) What is the kinetic energy of the electron after the collision?

16. An x-ray photon of frequency 1.5×10^{19} cycles/sec undergoes Compton scattering. Its new frequency is 1.2×10^{19} cycles/sec. What is the change in the kinetic energy of the electron it has struck?

17. A photon colliding with an electron at rest gives the electron an energy of 10,000 ev. Find the initial and final frequencies of the photon.

18. Give as many reasons as you can to support (a) the wave theory of light, (b) the particle theory of light.

19. Find the de Broglie wavelength of a 10-ton truck whose speed is 60 mi/hr.

20. Calculate the de Broglie wavelength of (a) an electron whose speed is 1×10^8 m/sec, and (b) an electron whose speed is 2×10^8 m/sec. Use relativistic formulas.

21. Calculate the de Broglie wavelength of a proton whose kinetic energy is 1 Mev $(= 10^6$ ev). This calculation may be made nonrelativistically.

22. (a) How could you experimentally distinguish between a photon whose momentum is 10^{-22} kg·m/sec and an electron whose momentum is 10^{-22} kg·m/sec? (b) What is the energy of each? (c) What is the wavelength of each?

23. How could you experimentally distinguish between an electromagnetic wave whose wavelength is 10^{-11} m and an electron whose de Broglie wavelength is 10^{-11} m?

24. What is the approximate momentum imparted to a proton initially at rest by a measurement which locates its position to 10^{-11} m?

25. (a) An electron is confined in a box 10^{-9} m in length. What is the uncertainty in its velocity? (b) A proton is confined in the same box. What is the uncertainty in its velocity?

26. The electron in a hydrogen atom may be thought of as confined to a region surrounding the proton at a radius of 5×10^{-11} m. (a) Use the uncertainty principle to calculate the momentum of the electron, and from this calculate its kinetic energy. (b) What keeps the electron in this region despite its kinetic energy?

27. Show that Planck's radiation law yields the Stefan-Boltzmann law by integrating $E_f \, df$ over the entire spectrum from $f = 0$ to $f = \infty$. To do this make use of the definite integral

$$\int_0^\infty \frac{x^3 \, dx}{e^x - 1} = \frac{\pi^4}{15}.$$

From this result obtain a formula for the constant b of Eq. (1–16).

4

The Atom

"Enclosed I send the first chapter of my paper on the constitution of atoms. I hope that the next chapters shall follow in a few weeks. In the latest time I have had good progress with my work, and hope to have succeeded in extending the considerations used to a number of different phenomena; such as the emission of line-spectra, magnetism, and possibly an indication of a theory of the constitution of crystalline structures. I have, however, some difficulties in keeping it all together at the same time, and shall be very glad to have some of it published as soon as possible on account of the accumulating literature on the subject. As besides the paper is getting rather long for publishing at one time in a periodical, I have thought it best to publish it in parts. Therefore I shall be very thankful if you kindly will communicate the present first chapter for me to the Philosphical Magazine.

"As you will see, the first chapter is mainly dealing with the problem of emission of line-spectra, considered from the point of view sketched in my former letter to you. I have tried to show that it, from such a point of view, seems possible to give a simple interpretation of the law of the spectrum of hydrogen, and that the calculation affords a close quantitative agreement with experiments... I hope that you will find that I have taken a reasonable point of view as to the delicate question of the simultaneous use of the old mechanics and of the new assumptions introduced by Planck's theory of radiation." (Letter to Rutherford, March 6, 1913.)

Niels Bohr (1885–1962)

By the beginning of this century a substantial body of evidence had been accumulated in support of the idea that the chemical elements consist of atoms. The nature of the atoms themselves, however, was still a mystery, although a significant clue had been discovered. This clue was the fact that electrons are constituents of atoms, which suggests that electrical forces are significant in atomic phenomena. J. J. Thomson, whose work had led to the identification of the electron, proposed in 1898 that atoms are spheres of positively charged matter containing embedded electrons, much as a fruit cake is studded with nuts. Thirteen years later an important experiment was performed whose results forced the abandonment of Thomson's apparently plausible model and its replacement by a model which seemed at the time to raise more questions than it answered. As we shall learn in this chapter, the search for a way of reconciling the experimental data on the atom with the laws of physics has profoundly affected our ideas on the structure of matter.

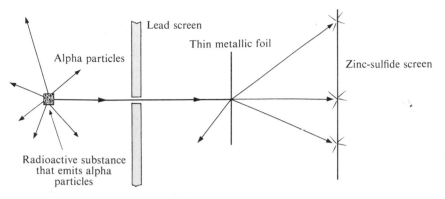

FIG. 4–1.   Diagram of the Rutherford experiment.

4–1 THE NUCLEAR ATOM

The most direct method for finding out what is inside a fruit cake is to simply plunge a finger into it. In essence this is the classic experiment performed in 1911 by Geiger and Marsden at the suggestion of Ernest Rutherford to find out what is inside an atom. The probes that they used were fast *alpha particles* spontaneously emitted by certain radioactive elements. For the present all we need know about alpha particles is that their mass is that of a helium atom and that they carry positive charges equal in magnitude to twice the charge on the electron e; we shall discuss their properties in more detail in Chapter 8. Geiger and Marsden placed a sample of an alpha-emitting substance behind a lead screen with a small hole in it, as shown in Fig. 4–1, so that a narrow beam of alpha particles is produced. On the other side of a thin metallic foil in the path of the beam they placed a movable zinc sulfide screen which gave off a flash of light when struck by an alpha particle, thus indicating the extent to which the alpha particles are scattered from their original direction of motion. They expected to find that most of the alpha particles go through the foil without being affected by it, with the remainder receiving only slight deflections. This anticipated behavior follows from the Thomson atomic model, in which the positive and negative electric charges within an atom are assumed to be spread more or less evenly throughout its volume (Fig. 4–2). If the Thomson model is correct, only weak electric forces would be exerted on alpha particles passing through a thin foil, and their momenta would be enough to carry them through with only minor departures from their original paths.

What Geiger and Marsden actually found was that, while most of the alpha particles indeed emerged unaffected from the foil, the others underwent deflections through very large angles, in some cases even being scattered in the backward direction (Fig. 4–1). Since alpha particles

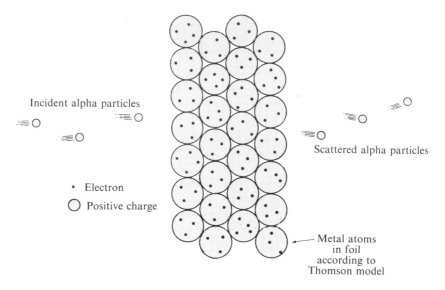

Incident alpha particles

Scattered alpha particles

• Electron
○ Positive charge

Metal atoms
in foil
according to
Thomson model

FIG. 4–2. According to the Thomson model, positive charge is spread uniformly throughout the volume of the atom with electrons embedded here and there. Only weak electric fields are present within atoms on the basis of this model, and it accordingly predicts very little deflection of alpha particles striking a thin foil. This prediction does not agree with experiment.

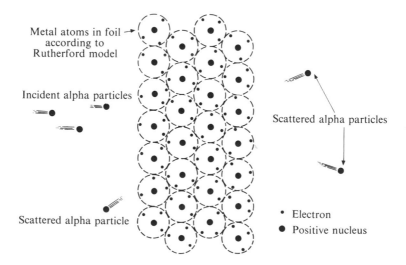

Metal atoms in foil
according to
Rutherford model

Incident alpha particles

Scattered alpha particles

Scattered alpha particle

• Electron
● Positive nucleus

FIG. 4–3. According to the Rutherford model, positive charge is concentrated in a tiny nucleus at the center of the atom with electrons some distance away. Strong electric fields are present within atoms on the basis of this model, and it accordingly predicts considerable deflection of alpha particles striking a thin foil. This prediction agrees with experiment.

are relatively heavy (almost 8000 times more massive than electrons) and have high initial speeds, it was clear that strong forces had to be exerted upon them to cause such marked deflections. To explain the results, Rutherford adopted the hypothesis that an atom is composed of a tiny *nucleus,* in which its positive charge and nearly all its mass are concentrated, with the electrons some distance away (Fig. 4–3). With the atom largely empty space, it is easy to see why most alpha particles proceed right through a thin foil. On the other hand, an alpha particle coming near a nucleus experiences a strong electric force, and is likely to be scattered through a large angle. (The atomic electrons, being very light, are readily knocked out of the way by alpha particles, while the situation is reversed for the nuclei, which are heavier than alpha particles.)

Rutherford was able to obtain a formula for the scattering of alpha particles by thin foils on the basis of his hypothesis that agreed with the experimental results. Rutherford is therefore credited with the "discovery" of the nucleus.

4–2 ELECTRON ORBITS: FORCE BALANCE

The picture of the atom that emerges from Rutherford's work is a tiny, massive nucleus with a positive charge surrounded by distant, light, negatively charged electrons. These electrons cannot be stationary, since the electrostatic force between them and the nucleus would attract them to the latter at once. If the electrons are in motion about the nucleus, however, dynamically stable orbits, like those of the planets about the sun, are possible.

Let us restrict ourselves for the time being to the hydrogen atom, which, having an atomic number of $z = 1$, is the simplest, with but a single proton as nucleus and one electron around it (Fig. 4–4). Then the centripetal force

$$F_c = \frac{mv^2}{r}$$

holding the electron in an orbit r from the nucleus must be provided by the electrostatic force

$$F_e = \frac{1}{4\pi\epsilon_0} \frac{e^2}{r^2}$$

between them, and

$$\frac{mv^2}{r} = \frac{1}{4\pi\epsilon_0} \frac{e^2}{r^2}. \tag{4-1}$$

The electron speed v is therefore related to its orbit radius r by the formula

$$v = \frac{e}{\sqrt{4\pi\epsilon_0 mr}}. \tag{4-2}$$

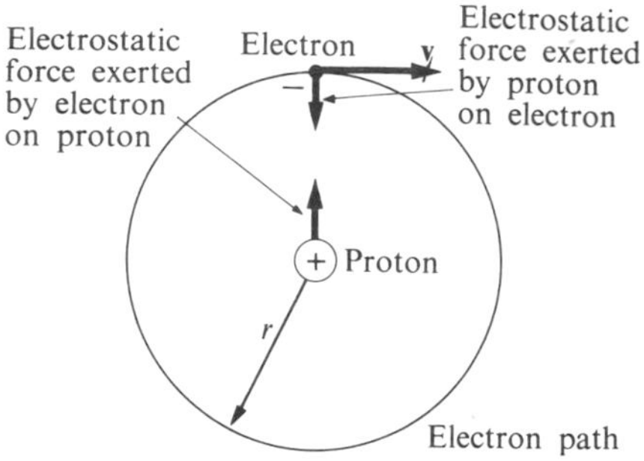

Electrostatic force exerted by electron on proton

Electron

Electrostatic force exerted by proton on electron

(+) Proton

r

Electron path

FIG. 4–4. The hydrogen atom consists of an electron circling a proton. The electrostatic force exerted by the proton on the electron provides the centripetal force required to hold it in a circular path. The proton is nearly 2000 times heavier than the electron, and so its motion under the influence of the electrostatic force the electron exerts is not significant.

The total energy E of the electron in a hydrogen atom is the sum of its kinetic energy

$$\mathrm{KE} = \tfrac{1}{2}mv^2$$

and its electrostatic potential energy

$$\mathrm{PE} = -\frac{1}{4\pi\epsilon_0}\frac{e^2}{r}.$$

(The latter formula follows from a calculation of the amount of work that would have to be done to bring the electron to an infinite distance from the proton; the potential energy here is a *negative* quantity since, being attracted to the proton, the electron cannot do work on anything outside the atom if it were to be released from its orbital motion.) Hence

$$E = \mathrm{KE} + \mathrm{PE} = \frac{mv^2}{2} - \frac{1}{4\pi\epsilon_0}\frac{e^2}{r}.$$

Substituting for the speed v from Eq. (4–2) yields

$$E = \frac{1}{4\pi\epsilon_0}\left(\frac{e^2}{2r} - \frac{e^2}{r}\right) = -\frac{1}{8\pi\epsilon_0}\frac{e^2}{r}. \qquad (4\text{–}3)$$

The total energy of an atomic electron is negative, implying that it is bound to the nucleus. If E were greater than zero, the electron would not be able to remain in a closed orbit about the nucleus.

Experiments indicate that 13.6 ev are required to separate a hydrogen atom into a proton and an electron, which means that its binding energy E is -13.6 ev. Since 13.6 ev $= 2.2 \times 10^{-18}$ joule, we can find the orbital radius of the electron in a hydrogen atom from Eq. (4–3). Since $1/4\pi\epsilon_0 = 9 \times 10^9$ n·m²/coul², this radius is

$$r = -\frac{1}{8\pi\epsilon_0}\frac{e^2}{E} = -\frac{9 \times 10^9 \ \mathrm{n{\cdot}m^2/coul^2} \times (1.6 \times 10^{-19} \ \mathrm{coul})^2}{2 \times (-2.2 \times 10^{-18} \ \mathrm{j})}$$

$$= 5.3 \times 10^{-11} \ \mathrm{m}.$$

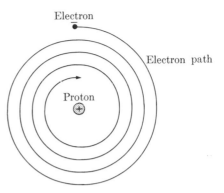

FIG. 4–5. According to electromagnetic theory, an accelerated electron radiates electromagnetic waves. The electron in a hydrogen atom, since it constantly experiences a centripetal acceleration, should therefore radiate energy and spiral into the nucleus.

The above analysis is dynamically flawless, but unfortunately it requires that the electron be continuously accelerated owing to its circular motion. And, according to electromagnetic theory, all accelerated electric charges must radiate electromagnetic waves. It is the acceleration experienced by a fast electron when it strikes a target that causes it to emit x-rays, for instance. An atomic electron circling its nucleus to keep from falling into it by electrostatic attraction therefore cannot (according to electromagnetic theory) help radiating away energy, thereby spiraling inward as r decreases until it is swallowed up by the nucleus (Fig. 4–5). Conventional physics is thus completely unable to account for the observed existence of stable atoms whose electrons remain at great distances from their respective nuclei, and we shall find that the once revolutionary ideas discussed in the previous chapter must be called upon if we are to understand the atom.

4–3 ATOMIC SPECTRA

We have already discussed the fact that heated solids emit radiation whose spectra are continuous, with all wavelengths present although in varying intensities. It happens that the observed features of blackbody radiation can be explained on the basis of the quantum theory without our having to consider the details of the radiation process itself; this implies that the nature of the heated solid is irrelevant to the process. Since the atoms of a solid are packed closely together, it is reasonable to suppose that their constituent electrons interact with one another. Hence, when the solid is heated to incandescence, any electron behavior characteristic of a particular atomic species becomes subordinate to the collective behavior of a great many interacting electrons.

At the other extreme, the atoms or molecules in a rarefied gas are so far apart on the average that they can be considered as not interacting with one another at all except during brief, relatively infrequent collisions. Here we would expect any emitted radiation to reflect the properties of the individual atoms or molecules that are present, an expectation

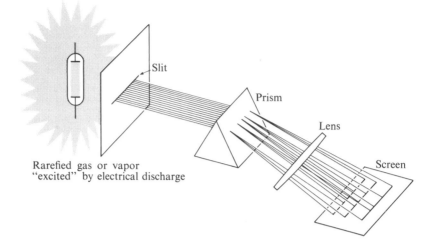

FIG. 4-6. The spectrum of a substance may be obtained by "exciting" a sample of its vapor in a tube by means of an electrical discharge and directing the light it gives off through a prism.

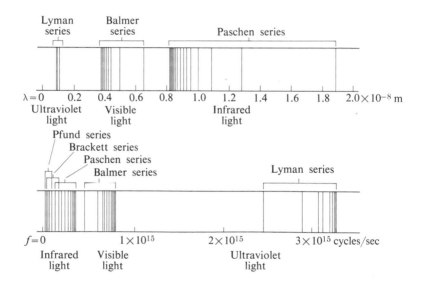

FIG. 4-7. The line spectrum of hydrogen. Note the presence of various series of spectral lines.

FIG. 4-8. A portion of the spectrum of the molecular gas NO showing the presence of bands rather than lines.

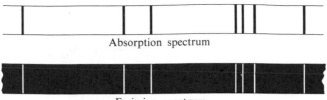

Absorption spectrum

Emission spectrum

FIG. 4-9. Portions of the emission (bright line) and absorption (dark line) spectra of an element.

that is fulfilled experimentally. When an atomic gas or vapor at some-what less than atmospheric pressure is "excited" by the passage of an electric current through it, light whose spectrum consists of a limited number of discrete wavelengths is emitted. A typical laboratory arrangement for observing such spectra is sketched in Fig. 4–6. Figure 4–7 shows the atomic spectrum of hydrogen; it is called a *line spectrum* from its appearance. Every element exhibits a unique line spectrum when a sample of it is suitably excited, and its presence in a substance of unknown composition can be ascertained by the appearance of its characteristic wavelengths in the spectrum of the substance.

When a molecular gas or vapor at low pressure is excited, "bands" consisting of many individual lines very close together are found in its spectrum (Fig. 4–8). The presence of bands in molecular spectra arises from interactions among the electrons of the atoms in each molecule; hence molecular spectra, like blackbody spectra, do not provide a great deal of insight into atomic structure.

It is worth noting that while unexcited gases (for instance, atomic hydrogen at room temperature) do not radiate their characteristic spectral lines, they *absorb* light of those particular wavelengths when white light is passed through them. In other words, the *absorption spectrum* of a gas is the same as its *emission spectrum*. Emission spectra consist of bright lines on a dark background, while absorption spectra consist of dark lines on a bright background (Fig. 4–9). The dark *Fraunhofer lines* in the solar spectrum (Fig. 4–10) occur because the luminous sun, which radiates approximately like a blackbody at 5800°K, is surrounded by an envelope of cooler gas that absorbs light of certain characteristic wavelengths.

The early spectroscopists found that the wavelengths present in atomic spectra fall into definite series, all of which can be expressed by similar

FIG. 4-10. The most prominent Fraunhofer lines in the visible part of the solar spectrum.

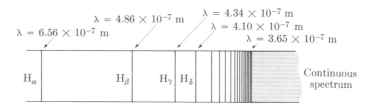

FIG. 4–11.   The Balmer series of spectral lines in atomic hydrogen.

simple formulas. The first spectral series was discovered by J. J. Balmer in 1885, who studied the wavelengths emitted by hydrogen in the visible part of its spectrum. As shown in Fig. 4–11, the *Balmer series* consists of a number of lines, the one with the longest wavelength being designated H_α, the next H_β, and so on. As the wavelength decreases, the lines are closer together and weaker in intensity until the *series limit* at a wavelength of 3.65×10^{-7} meter is reached, beyond which there are no more lines but only a faint continuous spectrum. Balmer's empirical formula for this series is

$$\text{Balmer:} \quad \frac{1}{\lambda} = R\left(\frac{1}{2^2} - \frac{1}{n^2}\right), \qquad n = 3, 4, 5, \ldots, \qquad (4\text{–}4)$$

where R, called the *Rydberg constant*, has the value $R = 1.097 \times 10^7 \, \text{m}^{-1}$. The H_α line corresponds to $n = 3$, the H_β line to $n = 4$, and so on. The series limit corresponds to $n = \infty$, so that its wavelength is $4/R$, in agreement with experiment.

The Balmer series only includes wavelengths in the visible part of the hydrogen spectrum. However, the spectral lines of hydrogen in the ultraviolet and infrared have been found to comprise four other series (Fig. 4–7). In the ultraviolet the *Lyman series* is specified by

$$\text{Lyman:} \quad \frac{1}{\lambda} = R\left(\frac{1}{1^2} - \frac{1}{n^2}\right), \qquad n = 2, 3, 4, \ldots, \qquad (4\text{–}5)$$

while in the infrared there are the three series

$$\text{Paschen:} \quad \frac{1}{\lambda} = R\left(\frac{1}{3^2} - \frac{1}{n^2}\right), \qquad n = 4, 5, 6, \ldots, \qquad (4\text{–}6)$$

$$\text{Brackett:} \quad \frac{1}{\lambda} = R\left(\frac{1}{4^2} - \frac{1}{n^2}\right), \qquad n = 5, 6, 7, \ldots, \qquad (4\text{–}7)$$

$$\text{Pfund:} \quad \frac{1}{\lambda} = R\left(\frac{1}{5^2} - \frac{1}{n^2}\right), \qquad n = 6, 7, 8, \ldots. \qquad (4\text{–}8)$$

In the above equations the value of R is the same as in Eq. (4–4).

The remarkable regularities in the hydrogen spectrum are not able to lead us directly to an elucidation of the structure of the hydrogen atom, but they do present a definite, quantitative test for any theory of this structure.

4-4 THE BOHR ATOM

The first theory of the atom to yield results that agree with our observations was proposed by Niels Bohr in 1913. By applying quantum ideas to atomic structure, Bohr arrived at a detailed model which, though subsequently replaced by an abstract description of greater accuracy and usefulness, remains today as the mental image most scientists have of the atom. Our derivation of the Bohr atom differs slightly from that originally given by Bohr, though all the results are identical.

As we saw, the ordinary laws of physics cannot account for the stability of the hydrogen atom, the simplest of all, whose electron must be whirling around the nucleus to keep from being pulled into it and yet must be radiating electromagnetic energy continuously. However, since other phenomena apparently impossible to understand—the photoelectric effect, for instance—found complete explanation in terms of quantum concepts, it is appropriate for us to inquire whether this might also be true for the atom. Let us begin by looking into the wave properties of the electron in a hydrogen atom. The de Broglie wavelength λ of any moving particle of mass m and speed v is

$$\lambda = \frac{h}{mv},$$

where h is Planck's constant. Now the speed of the electron in a hydrogen atom is, by Eq. (4–2),

$$v = \frac{e}{\sqrt{4\pi\epsilon_0 mr}},$$

and so its wavelength is

$$\lambda = \frac{h}{e}\sqrt{4\pi\epsilon_0 r/m}. \tag{4-9}$$

Substituting 5.3×10^{-11} meter for the radius r of the electron orbit, we find, again setting $1/4\pi\epsilon_0 = 9 \times 10^9$ n·m²/coul²,

$$\lambda = \frac{6.6 \times 10^{-34}\ \text{j·sec}}{1.6 \times 10^{-19}\ \text{coul}} \sqrt{\frac{5.3 \times 10^{-11}\ \text{m}}{9 \times 10^9\ \text{n·m}^2/\text{coul}^2 \times 9.1 \times 10^{-31}\ \text{kg}}}$$

$$= 33 \times 10^{-11}\ \text{m}.$$

This is a most exciting result, because the electron orbit is exactly

$$2\pi r = 33 \times 10^{-11}\ \text{m}$$

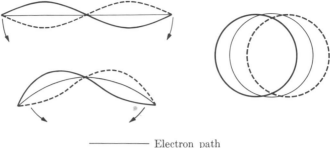

——————— Electron path
- - - - - - - - de Broglie electron wave

FIG. 4–12. The electron orbit in a hydrogen atom is exactly one de Broglie wavelength in circumference.

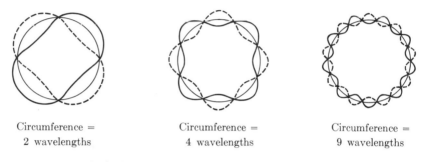

Circumference =   Circumference =   Circumference =
2 wavelengths   4 wavelengths   9 wavelengths

FIG. 4–13. Three possible modes of vibration of a wire loop.

in circumference. Therefore the orbit of the electron in a hydrogen atom corresponds to one complete electron wave joined on itself (Fig. 4–12).

The fact that the electron orbit in a hydrogen atom is one electron wavelength in circumference is just the clue we need to construct a theory of the atom. If we examine the vibrations of a wire loop, as in Fig. 4–13, we see that their wavelengths always fit an integral number of times into the loop's circumference, each wave joining smoothly with the next. In the absence of dissipative effects, such vibrations would persist indefinitely. Why are these the only vibrations possible in a wire loop? A fractional number of wavelengths cannot be fitted into the loop and still allow each wave to join smoothly with the next (Fig. 4–14); the result would be destructive interference as the waves travel around the loop, and the vibrations would die out rapidly. By considering the behavior of electron waves in the hydrogen atom analogous to the vibrations of a wire loop, then, we may postulate that

An electron can circle an atomic nucleus indefinitely without radiating energy if its orbit is an integral number of electron wavelengths in circumference.

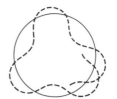

FIG. 4-14. Unless a whole (integral) number of wave-lengths fit into the wire loop, destructive interference causes the vibrations to die out rapidly.

[The electron speed must be that given by Eq. (4–2), of course, in order that the electrostatic attraction of the nucleus not pull the electron into it.]

The above postulate is the decisive one in our understanding of the atom. We note that it combines both the particle and wave characters of the electron into a single statement; while we can never observe these antithetical characters at the same time, they are inseparable in nature.

It is easy to express in a formula the condition that an integral number of electron wavelengths fit into the electron's orbit. The circumference of a circular orbit of radius r is $2\pi r$, and so the condition for orbit stability is

$$n\lambda = 2\pi r_n, \qquad n = 1, 2, 3, \ldots, \tag{4-10}$$

where r_n designates the radius of the orbit that contains n wavelengths. The quantity n is called the *quantum number* of the orbit. Substituting for λ the electron wavelength given by Eq. (4–9), we have

$$\frac{nh}{e} \sqrt{4\pi\epsilon_0 r_n/m} = 2\pi r_n,$$

and so the stable electron orbits are those whose radii are given by

$$r_n = \frac{\epsilon_0 n^2 h^2}{\pi m e^2}, \qquad n = 1, 2, 3, \ldots \tag{4-11}$$

The innermost orbit has the radius

$$r_1 = 5.3 \times 10^{-11} \text{ m},$$

in agreement with our previous calculation. The other radii are given in terms of r_1 by the formula

$$r_n = n^2 r_1,$$

so that the spacing between adjacent orbits increases progressively (Fig. 4–15).

4–5 ENERGY LEVELS AND SPECTRA

The energy of the electron is not the same in the various permitted orbits. The electron energy E_n is given in terms of the orbit radius r_n by Eq. (4–3) as

$$E_n = -\frac{1}{8\pi\epsilon_0} \frac{e^2}{r_n} \;;$$

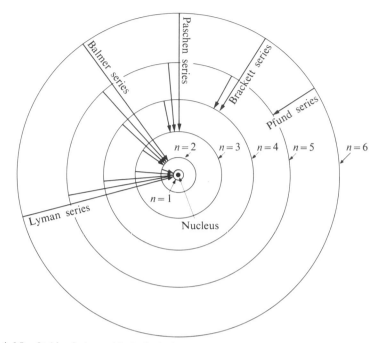

FIG. 4–15. Stable electron orbits in the hydrogen atom according to the Bohr theory. Some of the transitions between orbits that give rise to spectral lines are indicated.

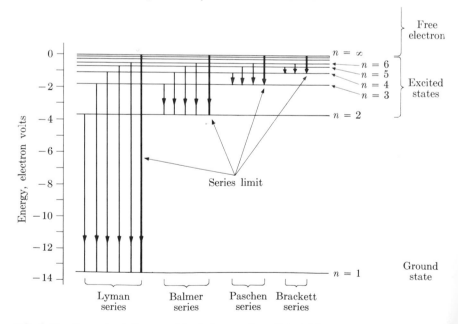

FIG. 4–16. Energy-level diagram of the hydrogen atom. Some of the transitions that give rise to spectral lines are indicated.

substituting for r_n from Eq. (4–11), we see that

$$E_n = -\frac{me^4}{8\epsilon_0^2 n^2 h^2}, \qquad n = 1, 2, 3, \ldots \tag{4-12}$$

The energies specified by Eq. (4–12), called the *energy levels* of the hydrogen atom, are plotted in Fig. 4–16. These levels are all negative, signifying that the electron does not have enough energy to escape from the atom. The lowest energy level E_1, corresponding to quantum number $n = 1$, is called the *ground state* of the atom; the higher levels E_2, E_3, E_4, . . . are called *excited* states. As the quantum number n increases, the corresponding energy E_n approaches closer and closer to zero; in the limit of $n = \infty$, $E_\infty = 0$ and the electron is no longer bound to the nucleus to form an atom. An energy greater than zero signifies an unbound electron, which, since it has no closed orbit that must satisfy quantum conditions, may have any positive energy value whatever.

While everything we have said thus far about the hydrogen atom seems perfectly reasonable, we must nevertheless find a way of directly confronting the equations we have developed with experiment. A particularly striking experimental result is that excited atoms emit line spectra. Can we show that these spectra are a natural consequence of our atomic model?

The presence of a series of definite, discrete energy levels in the hydrogen atom suggests a connection with line spectra. Let us tentatively assert that when an electron in an excited stated drops to a lower state, the difference in energy between the states is emitted as a single photon of light. Because electrons cannot, according to our model, exist in an atom except in certain specific energy levels, a rapid "jump" from one level to the other, with the difference in energy being given off all at once in a photon rather than in some more gradual manner, fits in well with this model. If the quantum number of the initial (higher-energy) state is n_i and the quantum number of the final (lower-energy) state is n_f, what we assert is that

Initial energy $-$ final energy $=$ quantum energy

$$E_i - E_f = hf, \tag{4-13}$$

where f is the frequency of the emitted photon.

The energies of the initial and final states of the electron in a hydrogen atom are, from Eq. (4–12), given by

$$\text{Initial energy} = E_i = -\frac{me^4}{8\epsilon_0^2 n_i^2 h^2},$$

$$\text{Final energy} = E_f = -\frac{me^4}{8\epsilon_0^2 n_f^2 h^2}.$$

Hence the energy difference between these states is

$$E_i - E_f = \frac{me^4}{8\epsilon_0^2 h^2}\left[\left(\frac{-1}{n_i^2}\right) - \left(\frac{-1}{n_f^2}\right)\right] = \frac{me^4}{8\epsilon_0^2 h^2}\left(\frac{1}{n_f^2} - \frac{1}{n_i^2}\right),$$

corresponding to a photon frequency f of

$$f = \frac{E_i - E_f}{h} = \frac{me^4}{8\epsilon_0^2 h^3}\left(\frac{1}{n_f^2} - \frac{1}{n_i^2}\right).$$

In terms of photon wavelength λ, since

$$\lambda = \frac{c}{f},$$

we have

$$\frac{1}{\lambda} = \frac{f}{c} = \frac{me^4}{8\epsilon_0^2 ch^3}\left(\frac{1}{n_f^2} - \frac{1}{n_i^2}\right). \tag{4–14}$$

Equation (4–14) predicts that the radiation emitted by excited hydrogen atom, are plotted in Fig. 4–16. These levels are all negative, wavelengths, furthermore, should fall into definite series depending upon the quantum number n_f of the final energy level of the electron. Since the initial quantum number n_i must always be greater than the final quantum number n_f in each case, in order that there be an excess of energy to be given off as a photon, the calculated formulas for the first five series are

$$n_f = 1: \quad \frac{1}{\lambda} = \frac{me^4}{8\epsilon_0^2 ch^3}\left(\frac{1}{1^2} - \frac{1}{n^2}\right), \quad n = 2, 3, 4, \ldots \tag{4–15}$$

$$n_f = 2: \quad \frac{1}{\lambda} = \frac{me^4}{8\epsilon_0^2 ch^3}\left(\frac{1}{2^2} - \frac{1}{n^2}\right), \quad n = 3, 4, 5, \ldots \tag{4–16}$$

$$n_f = 3: \quad \frac{1}{\lambda} = \frac{me^4}{8\epsilon_0^2 ch^3}\left(\frac{1}{3^2} - \frac{1}{n^2}\right), \quad n = 4, 5, 6, \ldots \tag{4–17}$$

$$n_f = 4: \quad \frac{1}{\lambda} = \frac{me^4}{8\epsilon_0^2 ch^3}\left(\frac{1}{4^2} - \frac{1}{n^2}\right), \quad n = 5, 6, 7, \ldots \tag{4–18}$$

$$n_f = 5: \quad \frac{1}{\lambda} = \frac{me^4}{8\epsilon_0^2 ch^3}\left(\frac{1}{5^2} - \frac{1}{n^2}\right), \quad n = 6, 7, 8, \ldots \tag{4–19}$$

These series agree in form with the empirical spectral series we discussed earlier, with the Lyman series, Eq. (4–5), corresponding to

$n_f = 1$, the Balmer series, Eq. (4–4), corresponding to $n_f = 2$, the Paschen series, Eq. (4–6), corresponding to $n_f = 3$, and so on!

Only one thing remains to be determined before we can consider our assertion that line spectra originate in electron transitions from high to low energy states as verified: does the constant term in Eqs. (4–15) through (4–19) have the same value as the Rydberg constant R of the empirical Eqs. (4–4) through (4–8)? Let us compute the value of this constant term. It is

$$\frac{me^4}{8\epsilon_0^2 ch^3}$$

$$= \frac{9.1 \times 10^{-31} \text{ kg} \times (1.6 \times 10^{-19} \text{ coul})^4}{8 \times (8.85 \times 10^{-12} \text{ coul}^2/\text{n·m}^2)^2 \times 3 \times 10^8 \text{ m/sec} \times (6.6 \times 10^{-34} \text{ j·sec})^3}$$

$$= 1.097 \times 10^7 \text{ m}^{-1},$$

which is the same as R. Our theory of the hydrogen atom, which is essentially that developed by Bohr in 1913, therefore agrees both qualitatively and quantitatively with experiment. Figure 4–16 shows schematically how spectral lines are related to atomic energy levels; Fig. 4–15 indicates the transitions in terms of Bohr orbits.

4–6 ATOMIC EXCITATION

There are two principal mechanisms by which an atom may be excited to an energy level above that of its ground state and thereby become capable of radiating. One of these mechanisms is a collision with another atom during which part of their kinetic energy is transformed into electron energy within either or both of the participating atoms. An atom excited in this way will then lose its excitation energy by emitting one or more photons in the course of returning to its ground state (Fig. 4–17). In an electric discharge in a rarefied gas, an electric field accelerates electrons and charged atoms and molecules (whose charge arises from either an excess or a deficiency in the electrons required to neutralize the positive charge of their nuclei) until their kinetic energies are sufficient to excite atoms they happen to collide with. A neon sign is a familiar example of how applying a strong electric field between electrodes in a gas-filled tube leads to the emission of the characteristic spectral radiation of that gas, which happens to be orange light in the case of neon.

Another excitation mechanism is the absorption by an atom of a photon of light whose energy is just the right amount to raise it to a higher energy level. A photon of wavelength 1.217×10^{-7} meter is emitted when a hydrogen atom in the $n = 2$ state drops to the $n = 1$ state; hence the absorption of a photon of wavelength 1.217×10^{-7} meter by a

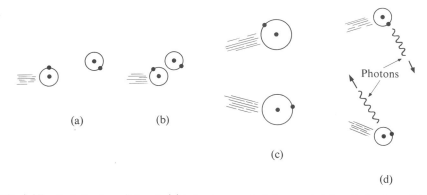

FIG. 4–17. Excitation by collision. In (a) both atoms are in their ground states. During the collision (b) some kinetic energy is transformed into excitation energy, and in (c) the atoms are both in excited states. In (d) they have returned to their ground states by emitting photons.

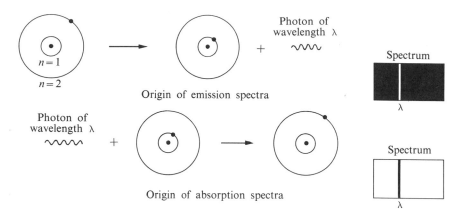

FIG. 4–18. The origins of emission and absorption spectra.

hydrogen atom initially in the $n = 1$ state will bring it up to the $n = 2$ state. This process explains the origin of absorption spectra (Fig. 4–18). When white light (in which all wavelengths are present) is passed through hydrogen gas, photons of those wavelengths that correspond to transitions between hydrogen energy levels are absorbed. The resulting excited hydrogen atoms reradiate their excitation energy almost at once, but these photons come off in random directions, not all in the same direction as in the original beam of white light (Fig. 4–19). The dark lines in an absorption spectrum are therefore never totally dark, but only appear so by contrast with the bright background of transmitted light. We would expect the absorption and emission spectra of a particular substance to be identical, which is what we find.

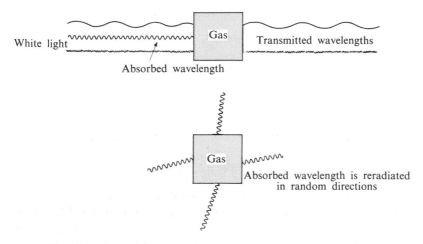

FIG. 4–19.   The dark lines in an absorption spectrum are never totally dark.

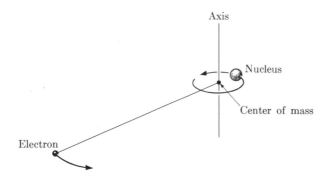

FIG. 4–20.   Electron and nucleus revolve about their common center of mass.

4–7 NUCLEAR MOTION

In our analysis of the hydrogen atom we have assumed that the nucleus is stationary while the electron revolves around it. This picture cannot be correct because the momentum of a hydrogen atom would have to vary continuously as its electron moves around the nucleus, contrary to the conservation of momentum in an isolated system. What actually happens is that nucleus and electron both revolve about their center of mass, which, because the nucleus is so much heavier than the electron, is very close to the nucleus (Fig. 4–20). In Appendix B we find that a system of this kind is exactly equivalent to a single particle of mass m' rotating about the position of the heavier particle, where m', the reduced mass of the system, is given by

$$ m' = \frac{m_1 m_2}{m_1 + m_2}. $$

To correct for nuclear motion in the hydrogen atom, then, all we have to do is imagine the electron replaced by a particle of mass m' revolving in the same orbit. The energy-level formula of Eq. (4–12) therefore becomes

$$E'_n = -\frac{m'e^4}{8\epsilon_0^2 n^2 h^2}, \qquad n = 1, 2, 3, \ldots, \qquad (4\text{-}19)$$

where, if m_e is the electron mass and m_n the nuclear mass,

$$m' = \frac{m_e m_n}{m_e + m_n}. \qquad (4\text{-}20)$$

In the ordinary hydrogen atom this change has the effect of shifting each of the energy levels upward by 0.06 percent, since each level becomes less negative than it would if the nucleus were stationary according to the formula

$$E'_n = \frac{m'}{m_e} E_n = \frac{m_n}{m_e + m_n} E_n$$

$$= \frac{1.672 \times 10^{-27} \text{ kg}}{1.673 \times 10^{-27} \text{ kg}} E_n$$

$$= 0.9994 E_n.$$

The correction for nuclear motion affects the hydrogen spectrum to a small but measurable amount. The Rydberg constant R for a stationary nucleus is, to six significant figures, $R = 1.09737 \times 10^{-7}$ m^{-1}, while it drops to 1.09678×10^{-7} meter when the reduced mass m' is substituted for the electron mass m in the formula for R. The latter figure is in better agreement with experiment than the former one.

A historically important application of the effect of nuclear motion was the discovery of a type of hydrogen atom whose nucleus has approximately twice the mass of the usual hydrogen nucleus. This variant of hydrogen is known as *deuterium*, or, less formally, as "heavy hydrogen"; about one hydrogen atom in 6000 is a deuterium atom. As we shall learn in Chapter 8, the deuterium nucleus consists of a neutron and a proton instead of a proton alone, which accounts for its greater mass. The spectrum of deuterium is identical with that of ordinary hydrogen except for an increase in the value of R from 1.09678×10^{-7} m^{-1} to 1.09707×10^{-7} m^{-1} owing to the larger reduced mass in the deuterium atom. The change means that the wavelengths of the spectral lines of deuterium are all slightly reduced. The H$_\alpha$ line, for example, which corresponds to a transition from $n = 3$ to $n = 2$, occurs at a wavelength of 6.56280×10^{-7} meter in ordinary hydrogen and at 6.56101×10^{-7} meter in deuterium. This minute difference was responsible for the discovery of deuterium in 1932.

PROBLEMS

1. In what ways do the Thomson and Rutherford atomic models agree? In what ways do they disagree?

2. In the Bohr theory of the hydrogen atom the electron is in constant motion. How is it possible for such an electron to have a *negative* amount of energy?

3. Calculate the speed of the electron in the innermost Bohr orbit of a hydrogen atom.

4. (a) Calculate the de Broglie wavelength of the earth. (b) What is the quantum number that characterizes the earth's orbit about the sun? (The earth's mass is 6.0×10^{24} kg, its orbital radius is 1.5×10^{11} m, and its orbital speed is 3×10^4 m/sec.)

5. Explain why the spectrum of hydrogen has many lines, although a hydrogen atom contains only one electron.

6. A beam of electrons whose energy is 13 ev is used to bombard gaseous hydrogen. What wavelengths of light will be emitted?

7. A beam of electrons is used to bombard gaseous hydrogen. What is the minimum energy in electron volts the electrons must have if the first line of the Balmer series, corresponding to a transition from the $n = 3$ state to the $n = 2$ state, is to be emitted?

8. What is the shortest wavelength present in the Paschen series of spectral lines?

9. What is the shortest wavelength present in the Brackett series?

10. A proton and an electron, both at rest initially, combine to form a hydrogen atom in the ground state. A single photon is emitted in this process. What is its wavelength?

11. How much energy (in joules and in electron volts) is required to remove the electron from a hydrogen atom when it is in the $n = 5$ state?

12. To what temperature must a hydrogen gas be heated if the average molecular kinetic energy is to equal the binding energy of the hydrogen atom?

13. When radiation with a continuous spectrum is passed through a volume of hydrogen gas whose atoms are all in their ground state, which spectral series will be present in the resulting absorption spectrum?

14. An electron spends about 10^{-8} sec in an excited state before it drops to a lower state by giving up energy in the form of a photon. How many revolutions does an electron in the $n = 2$ state of a hydrogen atom make before dropping to the $n = 1$ state?

15. (a) Find the wavelength of the photon emitted when a hydrogen atom goes from the state $n = 4$ to $n = 1$. (b) Find the momentum of this photon. (c) Find the recoil speed of the hydrogen atom after it has emitted this photon.

16. Calculate the average kinetic energy per molecule in a gas at room temperature (20°C), and show that this is much less than the energy required to raise a hydrogen atom from its ground state ($n = 1$) to its first excited state ($n = 2$). (See Appendix A.)

17. What kind of spectrum is observed in solar radiation during an eclipse?

18. How can the narrowness of spectral lines, which indicates that the responsible photons have very precisely defined energies, be reconciled with the version of the uncertainty principle that states that $\Delta E \, \Delta t \geq h$?

19. *Tritium* is a variety of hydrogen whose nuclei are approximately three times more massive than those of ordinary hydrogen. Show that the H_α line of tritium has a wavelength 2.4×10^{-10} m shorter than the H_α line of ordinary hydrogen.

5

Complex Atoms

"The periodic law has shown that our chemical individuals display a harmonic periodicity of properties dependent on their masses. Now natural science has long been accustomed to deal with periodicities observed in nature, to seize them with the vice of mathematical analysis, to submit them to the rasp of experiment. And these instruments of scientific thought would surely, long since, have mastered the problem connected with the chemical elements, were it not for a new feature which was brought to light by the periodic law, and which gave a peculiar and original character to the periodic function ... Not only are there no intermediate elements between silver, which gives AgCl, and cadmium, which gives $CdCl_2$, but, according to the very essence of the periodic law, there can be none; in fact a uniform curve would be inapplicable in such a case, as it would lead us to expect elements possessed of special properties at any point of the curve. The periods of the elements have thus a character very different from those which are so simply represented by geometers. They correspond to points, to numbers, to sudden changes of the masses, and not to a continuous evolution. In these sudden changes destitute of intermediate steps or positions, in the absence of elements intermediate between, say, silver and cadmium, or aluminium and silicon, we must recognize a problem to which no direct application of the analysis of the infinitely small can be made ... Having thus indicated a new mystery of Nature, which does not yet yield to rational conception, the periodic law, together with the revelations of spectrum analysis, have contributed to again revive an old but remarkably long-lived hope—that of discovering, if not by experiment, at least by a mental effort, the *primary matter*—which had its genesis in the minds of the Grecian philosophers, and has been transmitted, together with many other ideas of the classic period, to the heirs of their civilization."

Dmitri Mendeleev (1834–1907)

5–1 INTRODUCTION

The Bohr theory of the atom that we have sketched is indeed impressive in its agreement with experiment, but it has certain serious limitations. For one thing, the careful study of spectral lines shows that many of them actually consist of two or more separate lines that are close together, as in Fig. 5–1, something that the Bohr theory cannot account for. While correctly predicting the wavelengths of the spectral lines in hydrogen, which has but a single atomic electron, the Bohr theory fails when attempts are made to apply it to more complex atoms. Even in hydrogen it is not possible to calculate from the Bohr theory the relative probabilities of various transitions between energy levels, for instance whether it is more likely that an atom in the $n = 3$ state will go directly to the $n = 1$ state or instead first drop to the $n = 2$ state. (In other words, we cannot find from the Bohr theory which of the spectral lines of hydrogen will show up brightest in an emission spectrum and which will be faint.)

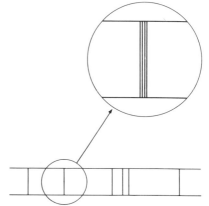

FIG. 5–1. Many spectral lines actually consist of several separate lines that are close together.

In cataloging these objections to the Bohr theory, the intent is not to detract from its eminence in the history of science, which is certainly secure, but instead to emphasize that a more general approach capable of wider application is necessary. Such an approach was developed in the 1920's by Schrödinger, Heisenberg, and others under the name of the *quantum theory of the atom*. Instead of trying to visualize an atomic electron as a kind of cross between a particle and a wave and thinking of it as occupying one of various possible orbits, the quantum theory of the atom avoids all reference to anything not capable of direct measurement and restricts itself only to such observable quantities as photon energies. In the Bohr theory we compute the radius of an electron orbit from a knowledge of its de Broglie wavelength, which depends upon the electron momentum; however, as we learned in Chapter 3, the position (and hence orbital radius) and momentum of an electron can never be simultaneously determined accurately, and so *even in principle* the Bohr theory cannot be subjected to an experimental test. The quantum theory sacrifices such intuitively accessible notions as that of electrons circling a nucleus like planets around the sun in favor of a wholly abstract mathematical formulation each of whose statements can be verified. Freed from the necessity of working in terms of a "model" of any kind, the quantum theory is able to tackle successfully a broad range of atomic problems.

5–2 QUANTUM NUMBERS

In the Bohr model of the hydrogen atom, the motion of the orbital electron is essentially one-dimensional. There the electron is regarded as being confined to a definite circular orbit, and the only quantity that changes as it revolves around its nucleus is its position on this fixed circle. A single quantum number is all that is required to specify the physical state of such an electron. In the more general quantum theory of the atom, the electrons have no such spatial restriction, and three different coordinates are needed to describe their motion. When the theory is worked out, three quantum numbers (one for each coordinate) turn out to be necessary instead of the single one of the Bohr theory. These are the *total quantum number n*, the *orbital quantum number l*, and the *magnetic quantum number* m_l.

The value of n is the chief factor that governs the total energy of an electron bound to a nucleus. The energy levels of a hydrogen atom are given by the formula

$$E_n = -\frac{me^4}{8\epsilon_0^2 h^2 n^2},\qquad (5\text{–}1)$$

the same result found by the Bohr theory.

A more novel conclusion of the quantum theory of the atom is the manner in which the angular momentum **L** of an electron is quantized. Angular momentum is a vector quantity, not a scalar quantity like energy, and has both magnitude and direction. The orbital quantum number l governs the magnitude L of an electron's angular momentum, which is restricted to

$$L = \sqrt{l(l+1)}\,\frac{h}{2\pi}.\qquad (5\text{–}2)$$

The orbital quantum number l can be 0 or any integer up to $n-1$; that is,

$$l = 0, 1, 2, \ldots, (n-1).$$

When $l = 0$, the angular momentum **L** $= 0$ also. Here is a significant difference from the Bohr theory, since an electron revolving in a circle must possess angular momentum, a requirement absent from the quantum theory.

The term *quantization* means that a certain quantity can have only certain discrete values. When we say that the direction of the angular momentum vector **L** is quantized, we mean that it can have only certain orientations in space. The space quantization of an atomic electron follows from the fact that, if such an electron has angular momentum, it behaves like an electric current loop and interacts with an external magnetic field in a manner similar to that of a bar magnet in such a field.

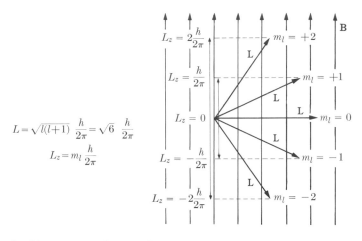

FIG. 5–2. Possible orientations of the angular momentum vector **L** in a magnetic field **B** when *l* = 2.

The larger the angular momentum **L**, the stronger the equivalent bar magnet. The potential energy of a bar magnet in a magnetic field **B** varies with the strength of the magnet and with its direction relative to **B**; the energy is least when the magnet is parallel to the field, and is most when it is antiparallel. The magnetic quantum number m_l determines the angle between **L** and **B** and hence governs the extent of the magnetic contribution to the total energy of the atom when it is in a magnetic field. Thus an atomic electron characterized by a certain value of m_l will assume a certain corresponding orientation of its angular momentum **L** relative to a magnetic field when placed in such a field. The orientation is specified in terms of the component of **L** parallel to the magnetic field. If we denote this component by L_z, as in Fig. 5–2, its possible values are

$$L_z = m_l \frac{h}{2\pi},$$ (5–3)

where m_l can be any integer from $-l$ through 0 to $+l$. That is,

$$m_l = -\,l,\, -\,(l-1),\, \ldots,\, 0,\, \ldots,\, +(l-1),\, +\,l.$$

An electron for which $l = 2$, for instance, could have a magnetic quantum number of -2, -1, 0, $+1$, or $+2$. When $m_l = 0$, $L_z = 0$ also, and **L** will be perpendicular to **B** whenever the atom is in a magnetic field. The angular momentum vector **L** can never be exactly parallel to **B**, even when $m_l = l$, since comparing Eqs. (5–2) and (5–3) shows that L_z is always less than L.

Because the magnetic quantum number m_l has several possible values for values of n other than $n = 1$, the presence of an external magnetic field splits the energy levels of a particular atom into two or more sub-

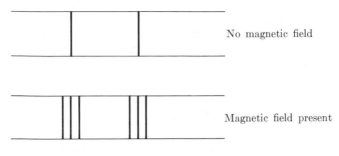

No magnetic field

Magnetic field present

FIG. 5-3. The Zeeman effect.

levels. The emission spectrum of an element in a magnetic field accordingly differs from its ordinary spectrum in that the spectral lines of the latter now have several components whose spacing varies with the magnetic flux density (Fig. 5-3). This phenomenon is known as the *Zeeman effect*. For example, a field of 1 weber/m² leads to an energy difference of 5.8×10^{-4} ev between adjacent sublevels of different m_l, which is observed as a wavelength difference between the split components of a given spectral line of order of magnitude 0.1%.

5-3 ELECTRON SPIN

We mentioned at the start of this chapter that one of the problems facing the atomic theorist is the observed splitting of many spectral lines into several components close together, as in Fig. 5-1. For example, the first line of the Balmer series of hydrogen (Fig. 4-11), which both theory and experiment place at a wavelength of 6.56280×10^{-7} m, actually consists of a pair of lines 0.00014×10^{-7} m apart. Furthermore, many cases of Zeeman splitting in magnetic fields cannot be explained solely in terms of the space quantization of angular momentum. In 1925 it was pointed out that the fine structure of spectral lines would occur if the electron behaved like a charged sphere spinning on its axis rather than like a single point charge; several years later the British physicist P.A.M. Dirac was able to show on the basis of a relativistic version of quantum theory that the electron *must* have the spin attributed to it. A spinning electron is effectively a tiny bar magnet, and it interacts with the magnetic field produced by its own orbital motion in an atom as well as with any magnetic fields originating outside the atom. Electron spin is described by the spin quantum number s, whose sole value is $s = \frac{1}{2}$. The magnitude of the intrinsic spin angular momentum of the electron is given by a formula similar to that describing the orbital angular momentum, namely

$$L_s = \sqrt{s(s+1)}\,\frac{h}{2\pi} = \frac{\sqrt{3}}{2}\frac{h}{2\pi}. \tag{5-4}$$

The space quantization of electron spin is described by the *spin magnetic quantum number*, m_s, whose values are $+\frac{1}{2}$ and $-\frac{1}{2}$. The component of spin angular momentum along a magnetic field is $m_s h/2\pi$, so that it can be either $+\frac{1}{2}h/2\pi$ *or* $-\frac{1}{2}h/2\pi$. Because the energy of an electron is different in each orientation of its spin owing to the presence of the magnetic field of its orbital motion, the various energy levels of an atom are split into sublevels, and the presence of these sublevels is responsible for the observed fine structure of spectral lines. (In its own frame of reference the electron "sees" the positively charged nucleus revolving around *it*, and it accordingly experiences a magnetic field like that present at the center of a loop of electric current.) The notion of electron spin is also able to explain those aspects of the Zeeman effect that cannot be understood on the basis of the space quantization of angular momentum alone.

5-4 THE EXCLUSION PRINCIPLE

The introduction of electron spin into the theory of the atom means that a total of four quantum numbers, n, l, m_l, and m_s, is required to describe each possible state of an atomic electron. All atoms other than hydrogen have more than one electron, and it is appropriate to inquire into their ground state configurations. Do all the electrons in a complex atom have the same sets of quantum numbers, or are the sets all different, or what? This question was answered by Wolfgang Pauli, whose *exclusion principle* states that

No two electrons in an atom can exist in the same quantum state.

That is, each electron in a complex atom must have a different set of the quantum numbers n, l, m_s, and m_l. The exclusion principle can be generalized to refer to the electrons in *any* small region of space, regardless of whether or not they constitute an atom. We shall make use of this generalization in the next chapter.

Pauli was led to the exclusion principle by a study of atomic spectra. It is possible to determine the quantum states of an atom empirically by analyzing its spectrum, since states with different quantum numbers differ in energy (even if only slightly) and the various wavelengths present correspond to transitions between these states. Pauli found several lines missing from the spectra of every element except hydrogen; the missing lines correspond to transitions to and from atomic states having certain sets of quantum numbers. Every one of these absent states has two or more electrons with identical sets of quantum numbers, a result that is expressed in the exclusion principle. Hydrogen, with a single electron, naturally has no absent states.

The Pauli exclusion principle is the final piece of information we require in order to understand atomic structure. Before we actually go into the electron configurations of the various atoms, however, it will be useful to review the periodic law of chemistry.

5–5 THE PERIODIC LAW

Certain elements resemble one another so closely in their chemical and physical properties that it is natural to think of them as forming a "family." Three particularly striking examples of such families are the *halogens,* the *inert gases,* and the *alkali metals.* The members of these groups of elements are listed in Table 5–1 with their atomic numbers.

TABLE 5–1

Three families of elements*

Halogens	*Inert gases*	*Alkali metals*
	(2) Helium	(3) Lithium
(9) Fluorine	(10) Neon	(11) Sodium
(17) Chlorine	(18) Argon	(19) Potassium
(35) Bromine	(36) Krypton	(37) Rubidium
(53) Iodine	(54) Xenon	(55) Cesium
(85) Astatine	(86) Radon	(87) Francium

* The atomic numbers of the elements are in parentheses.

The halogens are nonmetallic elements with a high degree of chemical activity. At room temperature fluorine and chlorine are gases, bromine a liquid, and iodine and astatine solids. The halogens have valences of −1, and form diatomic molecules in the vapor state. The inert gases, as their name suggests, are inactive chemically: they form virtually no compounds with other elements, and their atoms do not join together into molecules. The inert gases have valences of 0. The alkali metals, like the halogens, are chemically very active, but they are active as reducing agents rather than as oxidizing agents. They are soft, not very dense, and have low melting points (all but lithium are liquid above 100°C). The alkali metals have valences of +1.

A curious feature of the three groups listed above is that, while the atomic numbers of the member elements of each group bear no obvious relation to one another, each inert gas is preceded in atomic number by a halogen (except in the case of helium) and followed by an alkali metal. Thus fluorine, neon, and sodium have the atomic numbers 9, 10, and 11, respectively, a sequence that persists through astatine (85), radon (86), and francium (87). If we list *all* of the elements in order of their atomic number, *elements with similar properties recur at regular intervals.* This

TABLE 5–2

Periodic classification of the elements

Group→ Period↓	I	II											III	IV	V	VI	VII	0
1	1.0080 H 1																	4.0026 He 2
2	6.939 Li 3	9.012 Be 4											10.811 B 5	12.011 C 6	14.007 N 7	15.999 O 8	18.998 F 9	20.183 Ne 10
3	22.990 Na 11	24.31 Mg 12											26.98 Al 13	28.09 Si 14	30.97 P 15	32.06 S 16	35.45 Cl 17	39.95 Ar 18
4	39.10 K 19	40.08 Ca 20	44.96 Sc 21	47.90 Ti 22	50.94 V 23	52.00 Cr 24	54.94 Mn 25	55.85 Fe 26	58.93 Co 27	58.71 Ni 28	63.54 Cu 29	65.37 Zn 30	69.72 Ga 31	72.59 Ge 32	74.92 As 33	78.96 Se 34	79.91 Br 35	83.80 Kr 36
5	85.47 Rb 37	87.62 Sr 38	88.91 Y 39	91.22 Zr 40	92.91 Nb 41	95.94 Mo 42	(99) Tc 43	101.07 Ru 44	102.91 Rh 45	106.4 Pd 46	107.87 Ag 47	112.40 Cd 48	114.82 In 49	118.69 Sn 50	121.75 Sb 51	127.60 Te 52	126.9 I 53	131.30 Xe 54
6	132.91 Cs 55	137.34 Ba 56	* 57–71	178.49 Hf 72	180.95 Ta 73	183.85 W 74	186.2 Re 75	190.2 Os 76	192.2 Ir 77	195.09 Pt 78	196.97 Au 79	200.59 Hg 80	204.37 Tl 81	207.19 Pb 82	208.98 Bi 83	210 Po 84	(210) At 85	222 Rn 86
7	(223) Fr 87	226.05 Ra 88	† 89															

*Rare-earth metals	138.91 La 57	140.12 Ce 58	140.91 Pr 59	144.27 Nd 60	(147) Pm 61	150.35 Sm 62	151.96 Eu 63	157.25 Gd 64	158.92 Tb 65	162.50 Dy 66	164.93 Ho 67	167.26 Er 68	168.93 Tm 69	173.04 Yb 70	174.97 Lu 71
† Actinide metals	227 Ac 89	232.04 Th 90	231 Pa 91	238.03 U 92	(237) Np 93	(242) Pu 94	(243) Am 95	(245) Cm 96	(249) Bk 97	(249) Cf 98	(253) E 99	(255) Fm 100	(256) Mv 101	(253) No 102	(257) Lw 103

observation, first formulated in detail by Dmitri Mendeleev about 1869, is known as the *periodic law*.

A periodic table is a listing of the elements according to atomic number in a series of rows such that elements with similar properties form vertical columns. Table 5–2 is perhaps the most common form of periodic table; the number above the symbol of each element is its atomic weight, and the number below the symbol is its atomic number. The elements whose atomic weights appear in parentheses are radioactive and are not found in nature, but have been prepared in nuclear reactions. The atomic weight in each such case is the mass number of the longest-lived isotope of the element.

The columns in the periodic table are called *groups*. We recognize group I as the alkali metals plus hydrogen, group VII as the halogens, and group 0 as the inert gases. In addition to the elements forming the eight principal groups there are a number of *transition elements* falling between groups II and III. The transition elements are metals which share certain general properties: most are hard and brittle, have high melting points, exhibit several different valences, and form compounds that are paramagnetic. The rows in the periodic table are called *periods*. Each period starts with an active alkali metal and proceeds through less active metals to weakly active nonmetals to an active halogen and an inactive inert gas. The transition elements in each period may be very much alike; the rare earths and actinides are so much alike that they are usually considered as separate categories.

For nearly a century the periodic law has been a mainstay of the chemist by permitting him to predict the behavior of undiscovered elements and by providing a framework for organizing his knowledge. It is one of the triumphs of the quantum theory of the atom that it enables us to account for the periodic law in complete detail without invoking any new assumptions or postulates.

5–6 ATOMIC STRUCTURE

A physical system is stable when its energy is a minimum, which means that in the normal configuration of an atom its constituent electrons are present in the lowest energy states available to them. The electrons in an atom that share the same total quantum number n are said to occupy the same *shell*. This is a useful description, since such electrons average about the same distance from the nucleus and have comparable, though not identical, energies. Electrons in a given shell that also have the same orbital quantum number l are said to occupy the same *subshell*. In complex atoms the various subshells of the same shell vary in energy because electrons with different angular momenta pursue different orbits around the nucleus. (Strictly speaking, electrons do not

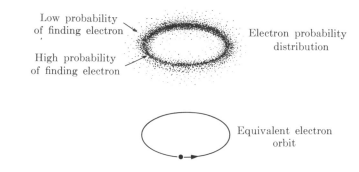

Low probability
of finding electron

High probability
of finding electron

Electron probability
distribution

Equivalent electron
orbit

FIG. 5-4. Electrons do not actually circle a nucleus in definite orbits, but it is convenient to think in terms of equivalent orbits that pass through regions of maximum electron probability.

move in specific orbits. However, the likelihood of their presence varies with position in such a way that a line drawn through the points of greatest probability may be thought of as an orbit. See Fig. 5-4.) In a complex atom the outer electrons are partially shielded from the full nuclear charge of $+Ze$ by the inner electrons, and are therefore less tightly bound than the latter. The extent of this shielding on a particular electron depends upon the shape of its orbit, which in turn depends upon its orbital quantum number l. When l is large, the corresponding orbit is more or less circular, while when it is small the orbit is pronouncedly elliptical. In an elliptical orbit an electron spends part of its time near the nucleus without being shielded by other electrons, as in Fig. 5-5, and its energy is correspondingly lower than that of an electron in the same shell whose orbit is circular. The general result, then, is that the energies of the various sublevels increase as l increases.

We shall now construct the periodic table of the elements with the help of the above considerations and the exclusion principle. Our procedure will be to investigate the status of a new electron added to an existing electronic structure. (Of course, the nuclear charge must also increase by $+e$ each time this is done.) In the simplest case we add an electron to a hydrogen atom ($Z = 1$) to give a helium atom ($Z = 2$). Both electrons in helium fall into the same $n = 1$ shell. Since $l = 0$ is the only value l can have when $n = 1$, both electrons have $l = m_l = 0$. The exclusion principle is not violated here since one electron can have the spin magnetic quantum number $m_s = +\frac{1}{2}$ while the other has $m_s = -\frac{1}{2}$. It is customary to describe this situation by saying that the electrons have *opposite spins*, that is, that they behave as though they rotate in opposite directions.

Because no more than two electrons can occupy the $n = 1$ shell, helium atoms have *closed shells*. The characteristic properties of closed shells and subshells are that the orbital and spin angular momenta of their con-

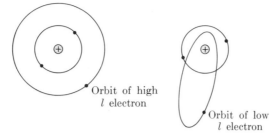

Orbit of high
l electron

Orbit of low
l electron

FIG. 5-5. An electron of low orbital quantum number spends part of the time near the nucleus, so its energy is lower than that of an electron in the same shell with a more circular orbit. The lower energy arises from the greater electrostatic force on an electron near the nucleus without shielding by other electrons, and it signifies tighter binding.

stituent electrons cancel out independently and that their effective electric charge distributions are perfectly symmetrical. As a result atoms with closed shells do not tend to interact with other atoms, which we know to be true of helium.

Lithium, with $Z = 3$, has one more electron than helium. There is no room left in the $n = 1$ shell, and so the additional electron goes into the $l = 0$ subshell of the $n = 2$ shell. The outer electron is relatively far from the nucleus in this shell, and is much less tightly bound. The chemical activity of lithium is a consequence of the low binding energy of this electron, which is readily lost to other atoms.

The next element, beryllium, has two electrons of opposite spin in the $l = 0$ subshell of the $n = 2$ shell. The nuclear charge is $+4e$, and so these outer electrons are more tightly held than the single outer electron in lithium; beryllium is accordingly less reactive than lithium.

Boron, with $Z = 5$, has an electron in the $l = 1$ subshell as well as two in the $l = 0$ one. The $l = 1$ subshell can contain a total of six electrons, corresponding to two electrons of opposite spin in the $m_l = +1$, $m_l = 0$, and $m_l = -1$ states. This subshell is closed (and the $n = 2$ shell is also closed) in neon, whose atomic number is 10. We therefore expect neon to be chemically inert, as indeed it is. Fluorine, the element just before neon in the periodic table, lacks but one electron of having a closed outer shell. Just as lithium tends to lose its single outermost electron in interacting with other elements, thereby leaving it with a closed shell configuration, fluorine tends to gain a single electron to close its outer shell. The very different behaviors of the alkali metals, the inert gases, and the halogens which we discussed in the previous section thus find explanation in terms of their respective atomic structures.

Table 5-3 shows the electron configurations of the 36 lightest elements. We note that the sequence of electron addition becomes irregular with

TABLE 5–3

Electron configurations
of the thirty-six lightest elements

Atomic number	Symbol	Element	Electron configuration of atom						
			$n=1$ $l=0$	$n=2$ $l=0$	$n=2$ $l=1$	$n=3$ $l=0$	$n=3$ $l=1$	$n=3$ $l=2$	$n=4$ $l=0, l=1, \ldots$
1	H	Hydrogen	1						
2	He	Helium	2						
3	Li	Lithium	2	1					
4	Be	Beryllium	2	2					
5	B	Boron	2	2	1				
6	C	Carbon	2	2	2				
7	N	Nitrogen	2	2	3				
8	O	Oxygen	2	2	4				
9	F	Fluorine	2	2	5				
10	Ne	Neon	2	2	6				
11	Na	Sodium	2	2	6	1			
12	Mg	Magnesium	2	2	6	2			
13	Al	Aluminum	2	2	6	2	1		
14	Si	Silicon	2	2	6	2	2		
15	P	Phosphorus	2	2	6	2	3		
16	S	Sulfur	2	2	6	2	4		

			1s	2s	2p	3s	3p	3d	4s	4p
17	Cl	Chlorine	2	2	6	2	5			
18	Ar	Argon	2	2	6	2	6			
19	K	Potassium	2	2	6	2	6		1	
20	Ca	Calcium	2	2	6	2	6		2	
21	Sc	Scandium	2	2	6	2	6	1	2	
22	Ti	Titanium	2	2	6	2	6	2	2	
23	V	Vanadium	2	2	6	2	6	3	2	
24	Cr	Chromium	2	2	6	2	6	5	1	
25	Mn	Manganese	2	2	6	2	6	5	2	
26	Fe	Iron	2	2	6	2	6	6	2	
27	Co	Cobalt	2	2	6	2	6	7	2	
28	Ni	Nickel	2	2	6	2	6	8	2	
29	Cu	Copper	2	2	6	2	6	10	1	
30	Zn	Zinc	2	2	6	2	6	10	2	
31	Ga	Gallium	2	2	6	2	6	10	2	1
32	Ge	Germanium	2	2	6	2	6	10	2	2
33	As	Arsenic	2	2	6	2	6	10	2	3
34	Se	Selenium	2	2	6	2	6	10	2	4
35	Br	Bromine	2	2	6	2	6	10	2	5
36	Kr	Krypton	2	2	6	2	6	10	2	6

potassium, which has an electron in its $n = 4$ shell even though its $n = 3$ shell is incomplete. The origin of this apparent anomaly is that electrons in the $n = 4$, $l = 0$ subshell have less energy, and therefore are more tightly bound, than those in the $n = 3$, $l = 2$ subshell, since the orbits of the former electrons pass closer to the nucleus than the more circular orbits of the latter electrons. The energy difference between the $n = 3$, $l = 2$ and $n = 4$, $l = 0$ subshells is actually quite small, as we can see from the configurations of chromium ($Z = 24$) and copper ($Z = 29$). In both of these elements an additional electron is present in the $n = 3$, $l = 2$ subshell at the expense of the $n = 4$, $l = 0$ subshell, leaving a vacancy in the latter that is filled in the succeeding element.

The ferromagnetic properties of iron, cobalt, and nickel are a consequence of their unfilled $n = 3$, $l = 2$ subshells; without violating the exclusion principle, the electrons in these subshells do *not* pair off, and their spins do not cancel out. In the $n = 3$, $l = 2$ subshell of iron, for instance, five of the six electrons have parallel spins, leaving each iron atom with a strongly magnetic character.

As we can see, the quantum theory of the atom is one of the most powerful and fruitful approaches yet devised for understanding the properties of matter.

PROBLEMS

1. Show that the electron structure of the inert gas xenon ($Z = 54$) contains only closed subshells.

2. Show that the electron structure of the alkali metal francium ($Z = 87$) has a single electron in its outermost shell.

3. How many electrons do the elements in group II of the periodic table have in their outermost shells?

4. Find the maximum number of electrons that can be present in each of the subshells of the $n = 4$ shell.

5. Show that the maximum number of electrons in a shell of given n is $2n^2$. (*Hint:* Multiply the average number of electrons in the subshells of the nth shell by the number of subshells in that shell.)

6. How many elements would there be if atoms with occupied electron shells up through $n = 6$ could exist?

7. The energy required to remove an electron from an atom is called its ionization energy. Account for the fact that the ionization energies of the elements of atomic numbers 20 through 29 are very nearly the same, although wide variations are found in the ionization energies of other sequences of elements.

8. Each quantum state of specific total, orbital, and magnetic quantum numbers is called an *orbital* by chemists. How many electrons can be present in each orbital of an atom?

9. How can you account for the fact that lithium and sodium exhibit similar chemical behavior?

10. How can you account for the fact that fluorine and chlorine exhibit similar chemical behavior?

11. Why do lithium atoms not form molecules with each other in the way that fluorine atoms do?

12. The ionization energies of Li, Na, K, Rb, and Cs are, respectively, 5.4, 5.1, 4.3, 4.2, and 3.9 ev. All are in group I of the periodic table. Account for the decrease in ionization energy with increasing atomic number.

6

Molecules and Ions

"We shall say that there is a chemical bond between two atoms or groups of atoms in case that the forces acting between them are such as to lead to the formation of an aggregate with sufficient stability to make it convenient for the chemist to consider it as an independent molecular species. With this definition we accept in the category of chemical bonds not only the directed valence bond of the organic chemist but also, for example, the bond between sodium cations and chloride anions in the sodium chloride crystal, those between the aluminum ion and the six surrounding water molecules in the hydrated aluminum ion in solution or in crystals, and even the weak bond that holds together the two O_2 molecules in O_4. In general we do not consider the weak van der Waals forces between molecules as leading to chemical-bond formation; but in exceptional cases, such as that of the O_4 molecule mentioned above, it may happen that these forces are strong enough to make it convenient to describe the corresponding intermolecular interaction as bond formation." *

Linus Pauling (1901–   )

Atoms are most stable when their electron shells are closed, and tend to gain or lose electrons in order to secure closed shells by joining with other atoms. Two principal binding mechanisms are known, *ionic* and *covalent*. In ionic binding, electrons are effectively transferred from one atom to another, while in covalent binding, two atoms may share one or more pairs of electrons in order to provide each with a closed outer electron shell. In some molecules one or the other mechanism predominates, while in others both participate to comparable extents in holding molecules together.

6–1 IONIC BINDING

Let us see how ionic binding functions in the case of NaCl, the sodium chloride "molecule" of ordinary table salt. (Salt crystals are aggregates of Na and Cl atoms which, although they do not pair off into individual molecules, do interact through ionic bonds whose nature can be most simply examined in terms of hypothetical NaCl molecules.) The work

---

* From Linus Pauling, *The Nature of the Chemical Bond*, 3rd ed., Cornell University Press, 1960.

that must be done to remove an electron from an atom is called the *ionization energy* of that atom. An atom with just one electron in its outer shell has a relatively low ionization energy which reflects the shielding effect of the inner electrons that cancel out most of the charge of the nucleus (Fig. 6–1). Such an atom tends to lose the outer electron and thereby assume an electron configuration composed solely of closed shells. At the other extreme, an atom lacking just one electron for the completion of its outer shell has a relatively high ionization energy which reflects the greater attractive force exerted by the nucleus when the nuclear charge is less effectively shielded by the intervening electrons. Thus sodium, with filled $n = 1$ and $n = 2$ shells and 1 electron in its $n = 3$ shell, has an ionization energy of 5.1 ev, while chlorine, also with filled $n = 1$ and $n = 2$ shells but with 7 electrons in its $n = 3$ shell, has an ionization energy of 13.0 ev.

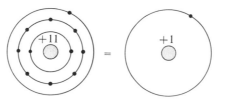

FIG. 6–1. Electron shielding. The presence of other electrons inside its orbit reduces the electrostatic attraction of the nucleus for an electron.

Another atomic property relevant to ionic binding is *electron affinity*. The electron affinity of an atom is the amount of energy released when it acquires an additional electron. Since this amount of work must be done to remove the additional electron after it is in place, electron affinity is a measure of the tendency of an atom to attract electrons in excess of its normal complement. As we would expect, atoms with high ionization energies have high electron affinities, while those with low ionization energies have low electron affinities. Atoms of the former kind, such as chlorine, tend to form negative ions by picking up electrons, and atoms of the latter kind, such as sodium, tend to form positive ions by losing electrons. An ionic bond between an atom of each kind occurs when one or more electrons pass from the atom with the low ionization energy to the one with the high electron affinity, since the resulting ions, both with complete outer shells, then attract each other electrostatically (Fig. 6–2). The bond is stable if more energy is needed to separate the ions than can be supplied by the return of the transferred electron or electrons.

As we said, the ionization energy of sodium is 5.1 ev, so that 5.1 ev of work is required to remove its single outer electron. The electron affinity of chlorine is 3.8 ev, so that 3.8 ev of energy is liberated when an electron is added to fill the vacancy in its outer shell. Transferring an electron from a sodium atom to a chlorine atom to form Na^+ and Cl^- ions there-

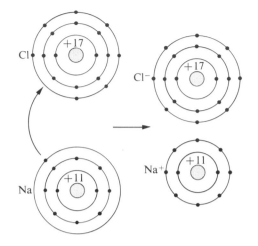

FIG. 6-2. Electron transfer in NaCl. The resulting ions attract each other electrostatically.

fore involves a net energy expenditure of 5.1 ev − 3.8 ev or 1.3 ev:

$$\text{Na} + \text{Cl} + 1.3\,\text{ev} \rightarrow \text{Na}^+ + \text{Cl}^-. \tag{6-1}$$

Now we consider the electrostatic potential energy of a charge of $+e$, corresponding to the Na^+ ion, that is a distance r from a charge of $-e$, corresponding to the Cl^- ion. This energy is

$$\text{PE} = -\frac{e^2}{4\pi\epsilon_0 r}\,; \tag{6-2}$$

the closer together the charges, the more negative their potential energy PE becomes. We recall that a system whose total energy is negative is stable, since work must be done on it from the outside to break it up by making its total energy positive. The PE of a system of a $+e$ and a $-e$ charge is -1.3 ev when the charges are 11×10^{-10} meter apart; hence the total energy of a Na^+ ion and a Cl^- ion is zero when the ions are separated by this distance. When the ions are still closer together, their total energy is negative, and external energy is needed to pull them apart. The actual separation of Na^+ and Cl^- in a NaCl molecule is 2.4×10^{-10} meter, so that this molecule is stable. The work that must be done to break a NaCl molecule into Na and Cl atoms is 4.2 ev; that is,

$$\text{NaCl} + 4.2\,\text{ev} \rightarrow \text{Na} + \text{Cl}. \tag{6-3}$$

There are two different repulsive effects that keep the Na^+ and Cl^- ions from being closer than 2.4×10^{-10} meter. Both effects arise because the electron structures of the ions mesh together when they are too near one another (Fig. 6-3). First, when such meshing occurs, the Na and Cl nuclei cease being shielded by their electrons and repel electrostati-

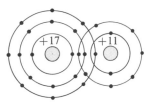

FIG. 6–3. Strong repulsive forces prevent atoms from coming so close together that their electron structures mesh.

cally. Second but more important, such meshing means that the Na^+ and Cl^- electrons no longer constitute separate atomic systems but instead constitute a single one. According to the Pauli exclusion principle, no two electrons in the same system can have identical sets of quantum numbers, and so some of the electrons must go into quantum states of higher energy than they otherwise would occupy. The result of both effects is that work must be done to push the ions together to make their electron structures overlap. The various attractive and repulsive forces cancel out at a separation of 2.4×10^{-10} meter, which is why the Na^+ and Cl^- ions in a NaCl molecule are this far apart.

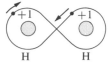

FIG. 6–4. Shared electrons spend more time on the average between their parent nuclei and therefore lead to a net attractive force.

H   H

6–2 COVALENT BINDING

In covalent binding the atoms composing a molecule share one or more pairs of electrons. Let us examine the origin of the covalent bond between the hydrogen atoms in a H_2 molecule, where ionic binding can play no part. If the shared electrons circulate around the H nuclei as shown in Fig. 6–4, they spend more time on the average between the nuclei than they do on the outside. The result is an effective negative charge between the positive H nuclei, and the attractive force this charge exerts on the nuclei is more than enough to counterbalance the direct repulsion between them. If the nuclei are too close together, however, their repulsion becomes dominant and the molecule is not stable. The balance between attractive and repulsive forces occurs at a separation of 7.42×10^{-11} meter where the total energy of the H_2 molecule is -4.5 ev. Hence 4.5 ev of work must be done to break a H_2 molecule into two H atoms:

$$H_2 + 4.5 \, ev \rightarrow H + H. \tag{6–4}$$

According to the Pauli exclusion principle, two electrons can jointly occupy the same region while in the lowest possible energy state when their spin quantum numbers are different, one having $s = +\frac{1}{2}$ and the other $s = -\frac{1}{2}$. In other words, the two electrons must have opposite

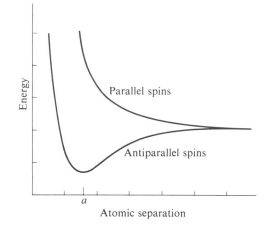

FIG. 6–5. The total energy of two hydrogen atoms plotted as a function of their separation. Binding is only possible when their electron spins are anti-parallel; the minimum of the lower curve corresponds to the H_2 molecule, which is a stable configuration since work must be done to move the H atoms closer together or farther apart than the distance a.

spins. If the spins were the same, the electrons would be restricted to spending most of their time near the far ends of the H atoms where they interact least. The result would be both a direct repulsion between the nuclei and a repulsion induced by the outward attraction of the electrons, making a stable molecule impossible (Fig. 6–5).

Carbon atoms tend to form four covalent bonds at the same time, since they have four electrons in their outer shells and these shells lack four electrons for completion. Various distributions of these bonds are possible, including bonds between adjacent carbon atoms. Chemists usually represent covalent bonds by using a dash for each shared pair of electrons. The structural diagrams of three common covalent molecules (methane, carbon dioxide, and acetylene) illustrate the different bonds in which carbon can participate to form a covalent molecule:

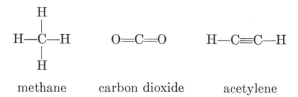

Carbon atoms are so versatile in forming covalent bonds with each other as well as with other atoms that literally millions of carbon compounds are known, some whose molecules contain tens of thousands of atoms. Such compounds were once thought to originate only in living things, and their study is accordingly known even today as organic chemistry.

6-3 POLAR MOLECULES

When atoms join together to form a molecule, their inner, complete electron shells undergo little change. In the limit of pure covalent binding, the atoms in a covalent molecule share their outer electrons evenly, and on the average remain neutral. In the opposite limit of pure ionic binding, however, electrons are transferred from certain atoms to other atoms in the molecule so that all have complete outer electron shells. The resulting molecule contains ions of opposite charge in different locations (Fig. 6-6). A molecule of this kind is called a *polar molecule,* while a molecule whose charge distribution is symmetric about its center and hence has no dipole moment is called a *nonpolar molecule.* Ionic molecules are polar, as we expect, but so are many covalent molecules: while their component atoms share electrons, the electrons may spend more time in the vicinity of some atoms than in the vicinity of others. Thus the covalent molecule HCl is polar because a chlorine atom has greater attraction for an electron than a hydrogen atom, while the covalent molecule H_2 is nonpolar because both hydrogen atoms have equal attractions for electrons.

FIG. 6-6. A polar molecule is one that behaves as though negatively charged at one end and positively charged at the other. The molecule as a whole is electrically neutral.

NaCl

The fact that polar molecules exist helps to explain a number of familiar phenomena. The behavior of compounds in solution is a good example. Water readily dissolves such compounds as salt and sugar, but cannot dissolve fats or oils. Gasoline readily dissolves fats and oils, but cannot dissolve salt or sugar. The key to these differences lies in the strongly polar nature of water molecules and the nonpolar nature of gasoline molecules. Water molecules tend to form aggregates under the influence of the electric forces between the ends of adjacent molecules, as shown in Fig. 6-7. Polar molecules of other substances, such as salt and sugar, can join in these aggregates, and are therefore easily dissolved by water. The nonpolar molecules of fats and oils, however, do not interact with water molecules. If samples of oil and water are mixed together, the attraction of water molecules for one another acts to squeeze out the oil molecules, and the mixture soon separates into layers of each substance. Fat and oil molecules dissolve only in liquids whose molecules are similar to theirs, which is why gasoline is a solvent for these compounds.

When an ionic compound dissolves in a strongly polar liquid like water, the forces attracting the ions to water molecules usually are

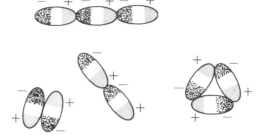

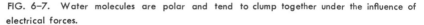

FIG. 6-7. Water molecules are polar and tend to clump together under the influence of electrical forces.

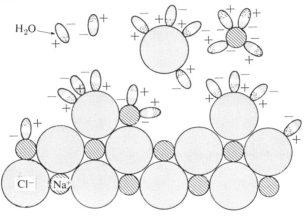

FIG. 6-8. The solution of solid NaCl.

greater than the forces holding the ions together. If a crystal of such a compound is placed in water, water molecules cluster around the crystal ions with their positive ends toward negative ions and their negative ends toward positive ions. The attraction of several water molecules is sufficiently great to extract the ion from the crystal lattice, and it moves into solution surrounded by water molecules (Fig. 6-8). The resulting solution contains ions rather than molecules of the dissolved compound. Substances that separate into free ions when dissolved in water are called *electrolytes* since they are able to conduct electric current by the migration of positive and negative ions. Figure 6-9 shows how a NaCl solution conducts electricity. All ionic compounds soluble in water and some polar covalent compounds, such as HCl, are electrolytes. Other covalent compounds, such as sugar, are nonelectrolytes even though they are soluble in water.

Since the outer electron structure of an ion may be very different from that of a neutral atom of the same species, it is not surprising that the ions of an element may have very different behavior from that of its atoms or molecules. Thus gaseous chlorine is greenish in color, has a

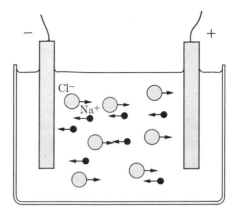

FIG. 6–9.   An electrolyte conducts electricity
by the motion of ions.

strong, irritating taste, and is very active chemically, while a solution of chlorine ions is colorless, has a mild, pleasant taste, and is only feebly active. The contrast in chemical activity, of course, follows from the incomplete outer electron shell in a chlorine atom while this shell is complete in a chlorine ion.

6–4 ELECTROLYSIS

Let us inquire into the effect of passing an electric current through a liquid containing ions. Because water itself may participate in the events that occur in a solution, for simplicity we shall consider molten NaCl rather than a NaCl solution; the quantitative results are the same in either case. When a current flows through a bath of molten NaCl metallic sodium is observed to deposit out at the negative electrode (or *cathode*) and gaseous chlorine to bubble up from the positive electrode (or *anode*). (At the temperature of molten NaCl, metallic sodium is also in the liquid state.) These results are not hard to understand in view of the presence of free Na^+ and Cl^- ions in the bath. The negative electrode attracts Na^+ ions and, when they arrive, neutralizes them by transferring an electron to each one:

$$Na^+ + e^- \rightarrow Na.$$

The resulting sodium atoms, unlike sodium ions, are not soluble and appear at this electrode as ordinary metallic sodium. The positive electrode at the same time attracts Cl^- ions and neutralizes them by absorbing an electron from each one:

$$Cl^- \rightarrow Cl + e^-.$$

The insoluble chlorine atoms are evolved as chlorine gas. The entire phenomenon is an example of *electrolysis*, the process by which free elements are liberated from a liquid by the passage of an electric current.

Over a century ago Michael Faraday discovered two quantitative laws that govern electrolysis. His experiments indicated that the mass of an element set free at either electrode is proportional to two quantities:

(1) The total quantity of charge Q that passed through the liquid;

(2) The atomic weight A of the element divided by the number v of electron charges ($+$ or $-$) on its ions.

The latter number, together with the sign of the charges, is called *valence* by chemists; the valence of Na is $+1$ since its ions, symbolized Na^+, each carry the charge $+e$, that of Cl is -1 since its ions, symbolized Cl^-, each carry the charge $-e$, and that of Ca is $+2$ since its ions, symbolized Ca^{++}, each carry the charge $+2e$. Denoting the atomic weight of an element by A and its valence (omitting the sign) by v, Faraday's laws together yield the formula

$$m = \frac{Q}{F}\frac{A}{v}, \tag{6-5}$$

where m is the deposited mass, Q the total charge transferred, and F a universal constant of proportionality. The value of F is 9.65×10^7 coul. Hence the passage of 9.65×10^7 coul through an appropriate liquid liberates one kmole of elements whose valence is ± 1 (that is, elements whose ions have the formulas X^+ or X^-), $\frac{1}{2}$ kmole of elements whose valence is ± 2 (ion formulas X^{++} or X^{--}), $\frac{1}{3}$ kmole of elements whose valence is ± 3 (ion formulas X^{+++} or X^{---}), and so forth. The quantity of charge 9.65×10^7 coul is known as the *faraday*.

Faraday's observations are consequences of the atomism of both matter and electricity. A kmole of every element contains the same number of individual atoms (Appendix A), which is Avogadro's number N_0. To convert an ion of a univalent element into the corresponding neutral atom requires the transfer of one electron; so to convert a whole kmole of that element requires the transfer of N_0 electrons. Thus the total charge involved is

$$Q = N_0 e = 6.025 \times 10^{26} \text{ atoms/kmole} \times 1.602 \times 10^{-19} \text{ coul/atom}$$

$$= 9.65 \times 10^7 \text{ coul/kmole} = 1 \text{ faraday}.$$

This result verifies our analysis of the electrolytic process. Evidently electrolysis provides a method for determining Avogadro's number, since the values of both the faraday, F, and the charge on the electron, e, can be measured in relatively direct experiments, while N_0 cannot be accurately determined by itself.

A higher valence than ± 1 means that more than one electron must be added or removed to neutralize each atom of the element. In this event less than a kmole will be liberated when 9.65×10^7 coul of charge passes through a liquid containing the ions of the element, an effect compensated for by the inclusion of the valence v in Eq. (6–5).

How much sodium will be produced if a current of 50 amp flows through a NaCl solution for 10 min? Since

$$Q = it,$$

the total charge transferred is

$$Q = 50 \frac{\text{coul}}{\text{sec}} \times 10 \text{ min} \times 60 \frac{\text{sec}}{\text{min}} = 3 \times 10^4 \text{ coul.}$$

A kmole of sodium contains 22.91 kg and the valence of sodium is +1. Substituting these figures in Eq. (6-5), we find that

$$m = \frac{Q}{F} \frac{A}{v} = \frac{3 \times 10^4 \text{ coul}}{9.65 \times 10^7 \text{ coul/kmole}} \times \frac{22.91 \text{ kg/kmole}}{1} = 7.13 \text{ kg}$$

of metallic sodium appear at the negative electrode.

When a solution of NaCl in water instead of molten NaCl undergoes electrolysis, chlorine appears at the anode and hydrogen at the cathode. Hydrogen rather than sodium is liberated because hydrogen atoms have the greater affinity for electrons. In effect, hydrogen and sodium in the solution compete for the available electrons, and the hydrogen wins. The left over Na^+ ions remain in the solution together with the OH^- (*hydroxide*) ions formed when H_2O molecules lose an H^+ ion each to produce the neutral hydrogen at the cathode. The bath acquires higher and higher concentrations of Na^+ and OH^- ions which can be recovered as solid NaOH by evaporating the solution. Thus the electrolysis of a water solution of inexpensive NaCl leads to the production of gaseous hydrogen and chlorine and the salt sodium hydroxide, all valuable industrial chemicals.

6-5 SPECIFIC HEATS OF GASES

A measure of the success of a scientific theory is how well it accounts for subsidiary aspects of the phenomenon it describes. As outlined in Appendix A, the kinetic theory of gases leads directly to the empirical ideal gas law, a notable achievement. But can this theory also yield information that agrees with experiment on such other properties of a gas as its specific heat, viscosity, and ability to conduct heat, all of which lie outside the domain of the ideal gas law?

Let us inquire into the specific heat of an ideal gas from the point of view of the kinetic theory of heat. The specific heat c of a substance is defined as the amount of heat that must be added to or removed from one kilogram of it to cause a change in temperature of 1°C. This simple statement cannot be used without qualification in the case of a gas, whose pressure or volume or both must change when the temperature changes. For example, a balloon expands when heat is added to it, and the gas within has performed work in moving the balloon wall outward against the elastic forces of the rubber. Some of the heat goes into this work as well as into raising the gas temperature. On the other hand, a gas sample confined in a rigid container cannot change in volume, and the pressure changes that occur when heat is added to or removed from it do not involve work. Thus the specific heat of a gas whose volume remains constant is always less than its specific heat under constant pressure.

In considering specific heat in terms of kinetic theory it is convenient to use the kmole as the unit of quantity instead of the kg, since kmoles of all substances contain the same number of molecules. We shall denote molar specific heat by the symbol C to distinguish it from specific heat per unit mass, c, and we shall use a subscript to indicate when a particular variable is held constant. Accordingly the specific heat of a gas per mole at constant volume is written C_V. At constant volume the amount of heat that must be added to 1 kmole of a gas to increase its temperature by dT is $C_V \, dT$. Since by hypothesis all of this heat goes into molecular energy, the average energy of each of the gas molecules increases from $\frac{3}{2}kT$ to $\frac{3}{2}k(T + dT)$, a change of $\frac{3}{2}kdT$. There are N_0 (Avogadro's number) molecules in the kmole of gas under consideration, and so

Heat added = increase in total molecular energy,

$$C_V \, dT = \tfrac{3}{2} N_0 k \, dT.$$

The quantity $N_0 k$ is equal to the universal gas constant R, and so, dividing the above equation through by dT, we obtain

$$C_V = \tfrac{3}{2} R. \tag{6-6}$$

To change the temperature of 1 kmole of an ideal gas by $1°K$ requires the addition or removal of $\frac{3}{2}R$ of heat. Since $R = 8.32 \times 10^3$ j/kmole·$°K$ or 1.99 kcal/kmole·$°K$, we expect that, in heat units,

$$C_V = \tfrac{3}{2} \times 1.99 \text{ kcal/kmole·}°K = 2.98 \text{ kcal/kmole·}°K.$$

How well is this prediction fulfilled by real gases? Table 6-1 is a list of the molar specific heats at constant volume for a number of common gases and vapors. Evidently theory and experiment agree reasonably well for gases composed of monatomic molecules, but gases with more complicated molecules all have higher heat capacities than $\frac{3}{2}R$. The agreement for monatomic gases indicates that we are certainly on the right track, but some relevant characteristic of polyatomic molecules has been neglected in deriving Eq. (6-6).

When we measure the temperature of a gas, what we actually determine is its average kinetic energy of translational motion. Since we assumed

TABLE 6-1

Molar specific heats at constant volume

Gas	C_V
Helium (He)	3.02 kcal/kmole·°K
Argon (Ar)	3.0
Mercury vapor (Hg)	3.0
Hydrogen (H$_2$)	4.8
Oxygen (O$_2$)	4.99
Carbon monoxide (CO)	5.02
Chlorine (Cl$_2$)	6.01
Sulfur dioxide (SO$_2$)	7.5
Ethane (C$_2$H$_6$)	10.3
Ether [(C$_2$H$_5$)$_2$O]	30.8

that a molecule is a minute structureless body whose mass can be regarded as being concentrated at a single point, this is the only kind of energy it can have. Adding $\frac{3}{2}R$ to 1 kmole of a monatomic gas causes its temperature to rise by almost exactly 1°K, but to increase the temperature of 1 kmole of a polyatomic gas by 1°K requires *more* than $\frac{3}{2}R$ of heat. Somewhere in the latter kind of gas is a repository of energy not found in a monatomic gas. It is not hard to imagine what such a repository might be. Even though we apparently are entitled to think of an *atom* as a point mass, we may not be justified in regarding a *combination of atoms* in so simple a manner. If a combination of atoms has extension in space, it can possess rotational kinetic energy even if the atoms are rigidly connected together. A diatomic molecule, for instance, is like a tiny dumbbell (Fig. 6–10), and because it has moments of inertia about

two perpendicular axes, can have kinetic energy associated with rotation about those axes. (Because we are considering atoms as point particles from the point of view of kinetic theory, a diatomic molecule has no moment of inertia about a line joining its two atoms. We shall see later in this chapter that this approximation is justified.) In the course of collision with the container walls and with other molecules, a diatomic molecule will be set into rotation, and molecular rotation therefore absorbs some of the total energy content of the gas.

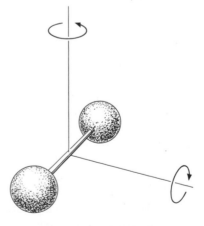

FIG. 6–10. A diatomic molecule.

As we said, a monatomic molecule can only possess translational kinetic energy. Now velocity is a vector quantity, and the three velocity components v_x, v_y, and v_z are needed to specify it completely. These components are mutually independent, with each unrelated to the others. A monatomic gas molecule may be said to have three *degrees of freedom*, corresponding to its ability to move in the x-, y-, and z-directions. By comparison, a particle restricted to the x-axis has only a single degree of freedom, while a particle fixed in place has no degrees of freedom. Because the pressure in a gas is the same in all directions, the molecules must be moving about in a random manner, and the three degrees of freedom share the available average molecular energy of $\frac{3}{2}kT$ equally. (The sharing is on an average basis too, of course.) Thus we can associate $\frac{1}{2}kT$ of energy with each one of the degrees of freedom of translational motion of a monatomic gas molecule:

$$\overline{\mathrm{KE}}_x = \tfrac{1}{2}kT, \qquad \overline{\mathrm{KE}}_y = \tfrac{1}{2}kT, \qquad \overline{\mathrm{KE}}_z = \tfrac{1}{2}kT.$$

The principle of *equipartition of energy* is a generalization of the above result. According to this principle, in an assembly of molecules the same amount of energy is associated with each degree of freedom. In this sense we may regard a degree of freedom as a mode of energy possession. A diatomic molecule therefore has a total of five degrees of freedom, three of translation and two of rotation; the latter correspond to the two independent rotational motions it can undergo. Each of these degrees of freedom represents an average energy of $\frac{1}{2}kT$ per molecule or a total of $\frac{5}{2}kT$. In an assembly of N_0 diatomic molecules, then, the total molecular energy content is $\frac{5}{2}N_0kT$ or $\frac{5}{2}RT$. By analogy with Eq. (6–6) we expect the molar specific heat for a diatomic gas to be

$$C_V = \tfrac{5}{2}R = 4.97 \ \mathrm{kcal/kmole \cdot {}^\circ K.} \tag{6–7}$$

From Table 6–1 we see that this figure is indeed in agreement with the specific heats of H_2, O_2, and CO.

While including molecular rotation does yield an adequate understanding of the specific heats of several diatomic molecules, it still predicts too low a value for chlorine, another diatomic molecule, and for such more complex molecules as sulfur dioxide, ethane, and ether even when an additional rotational degree of freedom is included to take into account their three-dimensional structures.

A hint as to the nature of the additional degrees of freedom possessed by complicated molecules can be obtained by studying the variation of C_V with temperature for a diatomic gas. Figure 6–11 shows this variation for hydrogen. The heat capacity of H_2 is approximately 5 kcal/kmole·°K only between about 250°K and 750°K, dropping to

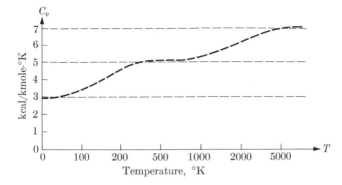

FIG. 6–11. Specific heat at constant volume of hydrogen gas vs. temperature. (A logarithmic scale is used for temperature.)

3 kcal/kmole·°K below 250°K and rising to 7 kcal/kmole·°K above 750°K. We can understand the reduced value of C_V at the lower temperatures by assuming that, for some reason, only translational motion is possible in a very cold gas. At the higher temperatures we can assume that the H_2 molecule vibrates like a pair of balls joined by a spring (Fig. 6–12). The vibration involves both kinetic and potential energies, providing two more degrees of freedom to the five of translation plus rotation. The heat capacity at high temperatures accordingly becomes $\frac{7}{2}R$ or 6.96 kcal/kmole·°K, almost exactly what is observed. Every link between the atoms in a molecule is therefore capable of adding two degrees of freedom and increasing the molar heat capacity by R; so the high specific heats of polyatomic gases have a plausible explanation.

We apparently have been able to understand the diverse specific heats of real gases by making no more than a reasonable extension of the

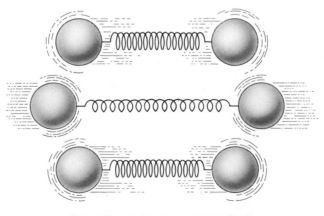

FIG. 6–12. A vibrating diatomic molecule.

simple kinetic-theory model of an ideal gas. However, while doing so, we have actually found something wholly contradictory to the fundamental assumption of the kinetic theory, which is that molecules can be treated as though they obey the ordinary laws of mechanics. If a diatomic molecule rotates and vibrates at high temperatures, why does it not do so at *all* temperatures? The behavior exhibited in Fig. 6–11 is really quite remarkable: kinetic theory can explain any *one* of the 3, 5, and 7 kcal/kmole·°K specific heat values of H_2, but not the occurrence of all three in different temperature ranges. We shall discover the reason for the nonclassical aspects of specific heats in the remainder of this chapter, where molecular physics is considered in the light of quantum theory. Adequate as the kinetic theory of gases is for many purposes, there is nevertheless a limit to its range of application.

6-6 MOLECULAR ROTATION

The energy levels of an atom arise from interactions between its electrons and nucleus and among the electrons themselves. Electronic energy levels are also present in a molecule but, owing to its composite structure, additional energy states representing rotation and vibration are possible. We have seen that the molecules of diatomic and polyatomic gases rotate and vibrate only at relatively high temperatures, although the molecules themselves are free to move about at all temperatures. This paradox is removed by quantum theory, which limits angular momenta and vibrational energies to certain specific values.

We shall begin by examining the rotation of a diatomic molecule about an axis that passes through its center of mass perpendicular to the line joining the two atoms (Fig. 6–13). The moment of inertia about this axis has the value

$$I = m'a^2 \qquad (6\text{-}8)$$

according to Appendix B, where

$$m' = \frac{m_1 m_2}{m_1 + m_2}$$

is the reduced mass. The molecule's angular momentum L when it rotates with the angular velocity ω is

$$L = I\omega.$$

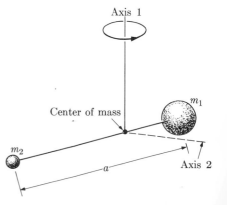

FIG. 6-13. An idealized model of a diatomic molecule.

The angular-momentum quantization of Eq. (5–2),

$$L = \sqrt{l(l+1)}\,\frac{h}{2\pi}, \qquad l = 0, 1, 2, 3, \ldots,$$

is not confined to atomic electrons alone, but turns out to be a general result that applies to all objects and systems that can possess angular momentum. Since Planck's constant h has the minute value of 6.63×10^{-34} j·sec, the quantization of angular momentum is imperceptible for macroscopic bodies. On a microscopic scale, however, this quantization dominates atomic and molecular phenomena. By combining the above expressions for L we find that the permitted angular velocities of a diatomic molecule are

$$\omega = \sqrt{l(l+1)}\,\frac{h}{2\pi I}, \qquad l = 0, 1, 2, 3, \ldots, \tag{6–9}$$

and the corresponding kinetic energies of rotation are

$$\mathrm{KE}_{\mathrm{rot}} = \tfrac{1}{2}I\omega^2 = l(l+1)\frac{h^2}{8\pi^2 I}, \qquad l = 0, 1, 2, 3, \ldots. \tag{6–10}$$

The lowest possible energy of rotation (other than zero) of a molecule whose moment of inertia is I may be found by substituting $l = 1$ in Eq. (6–10), which yields

$$\mathrm{KE}_{\mathrm{rot}}\,(\mathrm{min}) = \frac{h^2}{4\pi^2 I}. \tag{6–11}$$

For the H_2 molecule, where $m_1 = m_2 = 1.67 \times 10^{-27}$ kg and $a = 7.42 \times 10^{-11}$ m, the reduced mass is

$$m' = m^2/2m = \tfrac{1}{2}m = 8.4 \times 10^{-28}\ \mathrm{kg},$$

and so the moment of inertia is

$$I = m'a^2 = 8.4 \times 10^{-28}\ \mathrm{kg} \times (7.42 \times 10^{-11}\ \mathrm{m})^2$$

$$= 4.64 \times 10^{-48}\ \mathrm{kg \cdot m^2}.$$

Hence the least rotational kinetic energy a hydrogen molecule can have is

$$\mathrm{KE}_{\mathrm{rot}}\,(\mathrm{min}) = \frac{6.63 \times 10^{-34}\ \mathrm{j \cdot sec}}{4\pi^2 \times 4.64 \times 10^{-48}\ \mathrm{kg \cdot m^2}} = 2.39 \times 10^{-21}\ \mathrm{j}.$$

Rotation about a specific axis represents a degree of freedom, and each degree of freedom of a molecule in a gas has an average energy of $\tfrac{1}{2}kT$ associated with it. A diatomic molecule like H_2 can rotate about only two perpendicular axes (Fig. 6–13), since we are considering its atoms as point masses, and so its average total rotational energy is $2 \times \tfrac{1}{2}kT$ or

kT. When the temperature of a gas is so low that kT is much smaller than $KE_{rot}(min)$, virtually none of its molecules rotate, while if kT is much greater than $KE_{rot}(min)$, nearly all of them rotate. When kT is comparable to $KE_{rot}(min)$, some of the molecules rotate and some do not. Thus we expect a gradual transition in the specific heat of a gas with increasing temperature from the value representing pure translational motion to the value representing translation plus rotation. This expectation is fulfilled experimentally, as we can see in Fig. 6–11, where the molar specific heat of H_2 is plotted versus temperature.

The temperature at which kT and $KE_{rot}(min)$ are equal for a diatomic molecule is given by

$$T = \frac{h^2}{4\pi^2 kI}. \tag{6-12}$$

Since we have already computed $KE_{rot}(min)$ for H_2, we can find T by dividing it by Boltzmann's constant k:

$$T = \frac{2.39 \times 10^{-21} \text{ j}}{1.38 \times 10^{-23} \text{ j/}^\circ K} = 173^\circ K.$$

The precise value of this temperature is not to be taken too seriously, because the existence of definite rotational energy states means that the manner in which the available average kT of rotational energy per molecule is actually distributed among the molecules in a gas differs from that in which the average of $\frac{3}{2}kT$ of translational energy, not subject to any restriction, is distributed. The molecular speed distribution curves of Appendix A have quite different rotational counterparts. However, even though we cannot regard the T of Eq. (6–12) as representing, say, the temperature at which half the molecules in a gas rotate and half do not, it is usually not too far away from this temperature and is much easier to calculate. (An exact calculation shows that, at 173°K, a substantial fraction of the H_2 molecules is in the $l = 2$ rotational state and a very few are even in the $l = 3$ state.)

According to Eq. (6–11), the smaller the moment of inertia of a molecule, the greater its minimum kinetic energy of rotation and the greater T must be if it is to have an appreciable likelihood of rotating. Single atoms have such minute moments of inertia that temperatures of many thousands of degrees kelvin are required if rotation is to occur, which is why monatomic gases such as helium and argon have specific heats corresponding to translational motion only. The moment of inertia of a hydrogen atom, for example, is about 2.6×10^{-51} kg·m², while that of a hydrogen molecule is nearly 2000 times greater. As we saw, the molecules in an H_2 gas are rotating in appreciable numbers at 173°K, so that the comparable temperature in an atomic hydrogen gas would be about $350{,}000°\mathrm{K}$. For the same reason the rotation of a diatomic molecule about its axis of symmetry can be ignored compared with its rotation about axes perpendicular to this one.

6–7 MOLECULAR VIBRATION

The bonds that join atoms together to form molecules resist being stretched or compressed beyond their normal lengths, so to speak, and we therefore can crudely picture a molecule as a group of atoms attached together by tiny springs. If the usual molecular configuration is disturbed, say by a collision with another molecule or by the absorption of a suitable photon of electromagnetic energy, some or all of the atoms will vibrate. The vibrations of a diatomic molecule are easy to analyze because only a single bond is present. When the atoms of such a molecule are displaced by a small amount x from their normal separation, the restoring force that acts happens to be very close to the force $F = -kx$ of Hooke's law which is obeyed by ordinary springs. As described in Appendix C, a system of two particles of masses m_1 and m_2 joined by a spring of force constant k vibrates with the same frequency and total energy as a single particle of mass

$$m' = \frac{m_1 m_2}{m_1 + m_2}$$

attached to one end of the spring with the other end fixed in place (Fig. 6–14). The frequency of oscillation of a diatomic molecule is therefore

$$f = \frac{1}{2\pi}\sqrt{\frac{k}{m'}}. \tag{6–13}$$

One of the most significant results of quantum theory is that the total energy of a harmonic oscillator is limited to the values

$$E_v = (v + \tfrac{1}{2})hf, \qquad v = 0, 1, 2, 3, \ldots, \tag{6–14}$$

where f is the characteristic frequency of the oscillator and v, the *vibrational quantum number*, may be zero or any integer. The remarkable thing about this formula is not the quantization of energy it embodies, which is hardly surprising in view of the discussions of the previous few

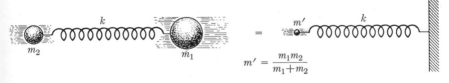

FIG. 6–14

chapters, but the fact that the lowest possible energy is $\frac{1}{2}hf$ and not zero. A harmonic oscillator cannot ever be completely at rest! The energy $\frac{1}{2}hf$ is called the *zero-point energy* E_0 of an oscillator with the natural frequency f, and, because Planck's constant h is so small, it is wholly imperceptible in the macroscopic world. The zero-point vibrations of a harmonic oscillator may be thought of as manifestations of the uncertainty principle: an oscillator at rest must have the precisely defined position $x = 0$ and momentum $mv = 0$, while the uncertainty principle does not permit our knowledge of both these quantities to be so exact. We thus have the curious fact that, at absolute zero, polyatomic molecules are still vibrating even though they can neither move nor rotate.

The force constant k of the bond in the H_2 molecule is found to be 5.73×10^2 n/m, and the reduced mass m' of the molecule is, as we found in the previous section, 8.4×10^{-28} kg. The natural frequency of vibration of the H_2 molecule is therefore

$$f = \frac{1}{2\pi} \sqrt{\frac{k}{m'}} = \frac{1}{2\pi} \sqrt{\frac{5.73 \times 10^2 \text{ n/m}}{8.4 \times 10^{-28} \text{ kg}}}$$

$$= 1.32 \times 10^{14} \text{ cycles/sec.}$$

The molecule's zero-point energy is

$$E_0 = \tfrac{1}{2}hf = \tfrac{1}{2} \times 6.63 \times 10^{-34} \text{ j·sec} \times 1.32 \times 10^{14} \text{ cycles/sec}$$

$$= 4.38 \times 10^{-20} \text{ j,}$$

and that of its first excited vibrational state ($v = 1$) is

$$E_1 = \tfrac{3}{2}hf = 13.13 \times 10^{-20} \text{ j.}$$

The zero-point oscillations do not contribute to the specific heat of gaseous hydrogen because they are always present, regardless of temperature. To go from the zero-point vibrational state to the first excited state requires an energy of

$$E_1 - E_0 = hf = 8.75 \times 10^{-20} \text{ j.}$$

Vibration involves two degrees of freedom, one for potential energy and one for kinetic energy, and as a result the average vibrational energy per molecule is $2 \times \frac{1}{2}kT$ or kT. As in the case of molecular rotation, we can equate kT with $E_1 - E_0$ to find what we might regard as a typical temperature at which an H_2 molecule will begin to vibrate in its $v = 1$ state. We find that

$$T = \frac{E_1 - E_0}{k} = \frac{8.75 \times 10^{-20} \text{ j}}{1.38 \times 10^{-23} \text{ j/°K}} = 6350°\text{K.}$$

An exact calculation shows that, at 6350°K, about half the molecules in gaseous hydrogen are indeed vibrating in excited states, but at this temperature many are in the $v = 2$, $v = 3$, and $v = 4$ states as well as in the $v = 1$ state.

It is evident that molecular vibration involves considerably higher energies than molecular rotation. Thus the energy of the first excited vibrational state in H_2 is 1.31×10^{-19} joule, while that of its lowest rotational state is 2.39×10^{-21} joule, 55 times smaller. A comparable ratio between vibrational and rotational energies holds for all molecules; hence any molecule that has enough energy to vibrate is certainly rotating as well. Much higher temperatures are required for the vibrational degrees of freedom in a polyatomic gas to absorb thermal energy than for the rotational ones, which is exactly the behavior found experimentally.

It is interesting to compare the frequencies of vibration and rotation in a particular molecule. As we saw, the natural frequency of vibration of the H_2 molecule is 1.32×10^{14} cycles/sec. From Eq. (6–9) we find that the angular velocity of the H_2 molecule when $l = 1$ is

$$\omega = \sqrt{l(l+1)}\, \frac{h}{2\pi I} = \frac{\sqrt{2} \times 6.63 \times 10^{-34}\ \text{j·sec}}{2\pi \times 4.64 \times 10^{-48}\ \text{kg·m}^2}$$

$$= 3.22 \times 10^{13}\ \text{rad/sec}.$$

The frequency of revolution f_r that corresponds to this angular velocity is

$$f_r = \frac{\omega}{2\pi} = 5.12 \times 10^{12}\, \frac{\text{rev}}{\text{sec}}\ ;$$

an H_2 molecule with $l = 1$ and $v = 1$ rotates 26 times for every vibration.

Our purpose in discussing molecular energy levels in some detail is threefold. First, the crucial role of quantization of energy and angular momentum is demonstrated to hold in the world of molecules as well as in the world of atoms. Second, the mysterious behavior of the specific heats of gases is clarified. And third, the theory of molecular energy levels is what has led to much of our detailed knowledge of molecular structure. Transitions among the various vibrational and rotational energy states in a molecule involve photons of various characteristic wavelengths, and the observed values of these wavelengths permit a great deal of information to be inferred. The exact arrangement of the atoms in a particular molecule can often be established from its rotational spectra, which are generally in the infrared region, since these spectra reveal the moments of inertia of the molecule about three axes. The strengths of the various interatomic bonds in a molecule can be determined from its vibrational spectra. Very complex molecules, such as most organic molecules, can be analyzed indirectly from their spectra. For instance, the bond between the O and H atoms in the OH^- group always has about the same force constant k regardless of the molecule in which OH^- appears, so that the appearance of the wavelengths characteristic of the vibrations of OH^- in the complete spectrum of a compound means that this group is present somewhere in its molecules. Spectroscopy has proved to be one of the most fruitful of our techniques for understanding the structure and behavior of molecules.

PROBLEMS

1. The ionization energy of potassium is 4.3 ev and the electron affinity of chlorine is 3.8 ev. (a) What is the net amount of energy required to form a K^+ and Cl^- ion pair from a pair of the same atoms? (b) Considering them as point charges, how close together must a K^+ and a Cl^- ion be if the total energy of the pair is to be zero?

2. The ionization energy of lithium is 5.4 ev and the electron affinity of bromide is 3.5 ev. (a) What is the net amount of energy required to form a Li^+ and Br^- ion pair from a pair of the same atoms? (b) Considering them as point charges, how close together must a Li^+ and a Br^- ion be if the total energy of the pair is to be zero?

3. Why are electrons much more readily liberated from lithium when it is irradiated with ultraviolet light than from fluorine?

4. Why do molecules increase in size with increasing temperature?

5. Why are Na atoms more active chemically than Na^+ ions?

6. Why are Cl atoms more active chemically than Cl^- ions?

7. Give an argument based on the Pauli principle to show that rare gas atoms (He, Ne, Ar, etc.) cannot participate in covalent bonds.

8. Give an argument based upon the Pauli principle to show that the molecule H_3 cannot exist.

9. Express the faraday in terms of amp·hr.

10. One faraday of charge is passed through a water bath. How much water is decomposed?

11. A current of 80 amp is passed through a chromium plating bath for 2 min. How much chromium is deposited? Chromium has the valence $+3$.

12. An 8-amp current is passed through a solution containing Cu^{++} ions. How long must the current flow to deposit 1 g of copper?

13. A current of 50 amp is passed through a bath of molten NaCl. How much sodium metal is produced per day?

14. A spoon whose surface area is 140 cm^2 is to be plated with a layer of silver 0.01 mm thick. How long should it stay in a plating bath in which a current of 1.2 amp is maintained? The density of silver is 1.05×10^4 kg/m^3, and its valence is $+1$.

15. A steel sheet 3 ft long and 2 ft wide is to be plated with 0.010 in. of copper. How long should it remain in an electroplating bath that contains Cu^{++} ions in which the current is 100 amp? The density of copper is 8.89×10^3 kg/m^3.

16. Commercial aluminum is refined electrolytically from a solution of aluminum oxide, Al_2O_3, in molten cryolite. The container, a carbon-lined iron box, is the cathode, and the anodes are carbon. The passage of current through the cell provides enough heat to keep the electrolyte at about 1000°C, at which temperature both the cryolite and the deposited aluminum are molten. Carbon dioxide is produced at the anodes. A typical potential difference in such a cell is 5 volts. If the price of electric power is $0.001/kw·hr, what is the cost of the electricity used in refining a ton of aluminum? Aluminum has the valence +3.

17. The Hg and Cl atoms in a certain HgCl molecule have masses of 3.32×10^{-25} kg and 5.81×10^{-26} kg respectively. The molecule emits a photon of wavelength 0.044 m when it undergoes a rotational transition from $l = 1$ to $l = 0$. What is the interatomic distance in this molecule?

18. The moment of inertia of a diatomic molecule increases as its rotational quantum number l increases. Compare this behavior with the equatorial bulge of the earth.

19. The dissociation energy of hydrogen is 4.5 ev (that is, 4.5 ev of work is required to separate the two hydrogen atoms that make up a hydrogen molecule). Assume that the hydrogen molecule behaves exactly like a harmonic oscillator and find the vibrational quantum number corresponding to a vibrational energy as close as possible to the dissociation energy.

20. A pendulum has a zero-point energy of 2.07×10^{-15} ev. Find its length.

21. The H and Cl atoms in a certain HCl molecule have masses of 1.66×10^{-27} kg and 5.81×10^{-26} kg respectively. If the separation of the atoms is 1.29×10^{-10} m, find the number of rotations per second made by the molecule when it is in the $l = 1$ rotational state.

22. The force constant of the bond in the HCl molecule is 470 n/m. (a) Find its natural frequency of oscillation. (b) Is it likely that the molecule will be vibrating in the $v = 1$ state at room temperature?

7

The Solid State

"Let us sum up the advantages of the metals. In the first place, they are useful to the physician, for they furnish liberally the ingredients for medicines, by which wounds and ulcers are cured, and even plagues; so that certainly if there were no other reasons why we should explore the depths of the earth, we should for the sake of medicine alone dig in the mines. Again, the metals are of use to painters, because they yield pigments which, when united with the painter's slip, are injured less than others by the moisture from without. Further, mining is useful to the architects, for thus is found marble, which is suitable not only for strengthening large buildings, but also for decoration. It is, moreover, helpful to those whose ambition urges them toward immortal glory, because it yields metals from which are made coins, statues, and other monuments, which, next to literary records, give men in a sense immortality. The metals are useful to merchants with very great cause, for, as I have stated elsewhere, the use of money which is made from metals is much more convenient to mankind than the old system of exchange of commodities. In short, to whom are the metals not of use? In very truth, even the works of art, elegant, embellished, elaborate, useful, are fashioned in various shapes by the artist from the metals gold, silver, brass, lead, and iron. How few artists could make anything that is beautiful and perfect without using metals? Even if tools of iron or brass were not used, we could not make tools of wood and stone without the help of metal. From all these examples are evident the benefits and advantages derived from metals. We should not have had these at all unless the science of mining and metallurgy had been discovered and handed down to us. Who then does not understand how highly useful they are, nay rather, how necessary to the human race? In a word, man could not do without the mining industry, nor did Divine Providence will that he should."

Georgius Agricola (1494–1555)

The same physical processes that can tie several atoms together into a molecule can also tie an unlimited number of them together into a solid. Other binding mechanisms are found in certain solids which confer upon them such remarkable properties as the ability to conduct electric current readily. While only an infinitesimal part of the universe is in the solid state, solids constitute much of the physical world of our experience, and modern technology is to a large extent based upon the exploitation of the unique characteristics of various solid materials.

7–1 THE STRUCTURE OF SOLIDS

Most solids are crystalline in nature, with their constituent atoms or molecules arranged in regular, repeated patterns. A crystal is thus characterized by the presence of *long-range order* in its structure. Other solids lack the definite arrangements of atoms and molecules so conspicuous in crystals, and may be regarded in a sense as supercooled liquids whose stiffness is due to an exaggerated viscosity. Examples of such *amorphous* ("without form") solids are pitch, glass, and many plastics. The structures of amorphous solids exhibit *short-range order* only. Some substances, for instance B_2O_3, can exist in either crystalline or amorphous forms. In both cases each boron atom is surrounded by three larger oxygen atoms, which represents a short-range order. In a B_2O_3 crystal a long-range order is also present, as shown in a two-dimensional representation in Fig. 7–1, while amorphous B_2O_3, a glassy material, lacks this additional regularity. The lack of long-range order in amorphous solids means that the various bonds vary in strength. When an amorphous solid is heated, the weakest bonds are ruptured at lower temperatures than the others and the solid softens gradually, while in a crystalline solid the bonds break simultaneously, and melting is a sudden process.

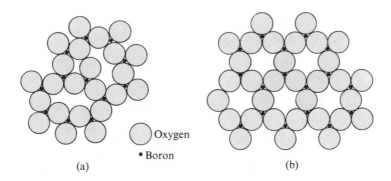

Oxygen
• Boron
(a)   (b)

FIG. 7–1. Structure of B_2O_3. (a) Amorphous B_2O_3, a glass, has short-range order only in its structure. (b) Crystalline B_2O_3 has long-range order as well.

How is it possible to determine the arrangement of atoms in a solid when atomic dimensions are so small? To appreciate this problem, let us calculate the spacing between adjacent atoms in rock salt, which is the crystalline form of NaCl. The density of rock salt is 2.17×10^3 kg/m, and the atomic weights of sodium and chlorine are respectively 22.99 and 35.46. The molecular weight of NaCl is therefore $22.99 + 35.46$, or 58.45, and each molecule contributes two atoms to the crystal lattice. The number of atoms per unit volume in the crystal is, since Avogadro's number is the number of molecules per kmole,

$$\text{Atomic density} = 2\,\frac{\text{atoms}}{\text{molecule}} \times 6.025 \times 10^{26}\,\frac{\text{molecules}}{\text{kmole}}$$

$$\times \frac{1}{58.45\,\text{kg/kmole}} \times 2.17 \times 10^{3}\,\frac{\text{kg}}{\text{m}^{3}}$$

$$= 4.47 \times 10^{28}\,\text{atoms/m}^{3}.$$

Let us suppose that each atom occupies a cube a on a side so that the volume it preempts in the crystal is a^3. Therefore

$$4.47 \times 10^{28}\,\text{atoms} \times a^{3} = 1\,\text{m}^{3}$$

$$a = 2.82 \times 10^{-10}\,\text{m}.$$

This minute distance represents the spacing between adjacent atoms in a rock-salt crystal.

The atomic structure of a crystal is on too fine a scale to be resolved using visible light, since the shortest wavelengths the eye responds to exceed a by a factor of 10^3. X-rays, however, can be produced with wavelengths well under 10^{-10} meter, which makes them ideal for investigating crystal structures. Several techniques have been developed for this purpose, and we shall examine an especially straightforward one devised in 1912 by W. L. Bragg. More recently methods employing electron and neutron beams of appropriate de Broglie wavelength have come into use in studying solids, and, in fact, it was the diffraction of electrons by crystals that provided the first experimental proof of de Broglie's hypothesis.

Figure 7–2 shows a beam of x-rays directed at the face of a crystal. Most of them go right through without interacting, but a few are scattered by atoms in their paths. The conditions for a scattered ray to emerge are the same as those for the reflection of light by a mirror: the scattered and incident rays must lie in a plane perpendicular to the crystal face, and the angle of incidence must equal the angle of scattering. However, x-rays can be scattered by atoms in the interior of the crystal as well as at its surface. Unless all of the x-ray waves scattered in a given direction are in phase ("in step"), destructive interference will result and they will cancel one another out. We can derive the condition for constructive interference by considering the scattering of a beam of x-rays from two successive layers of atoms a distance a apart, as in Fig. 7–3. If the beam is incident at the angle θ, both scattered rays also leave at this angle. Ray II must travel the additional distance relative to ray I of $a \sin \theta$ in order to reach its scattering atom, and then a further addi-

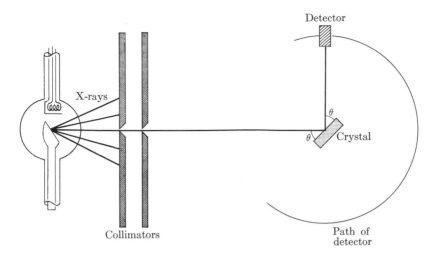

FIG. 7–2. X-ray scattering by a crystal. The angles of incidence and "reflection" are equal.

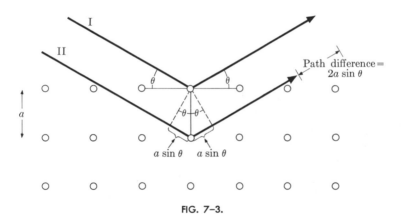

FIG. 7–3.

tional distance of $a \sin \theta$ to emerge from the crystal. Thus ray II travels $2a \sin \theta$ farther than ray I in the course of being scattered. If this path difference is a whole number of x-ray wavelengths, the two rays will reinforce each other. The condition for constructive interference is therefore

Path difference = whole number of wavelengths

$$2a \sin \theta = n\lambda, \qquad n = 1, 2, 3, \ldots \qquad (7\text{–}1)$$

By directing a beam of x-rays of known wavelength at a crystal and varying the angle of incidence in order to determine the angles at which scattering occurs, the spacing a between adjacent layers of atoms can

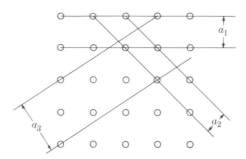

FIG. 7–4.   The atoms in a crystal fall into several series of parallel layers.

be ascertained. Of course, the atoms fall into more than one series of parallel layers, as in Fig. 7–4, and a different spacing will be obtained for each of them.

7–2 IONIC AND COVALENT BONDS

The ionic and covalent bonds between atoms that are responsible for the formation of molecules also act to hold many crystalline solids together. Figure 7–5 shows the structure of a NaCl crystal; the small spheres represent Na^+ ions and the large spheres Cl^- ions. Each ion behaves essentially like a point charge, and thus tends to attract to itself as many ions of opposite sign as can fit around it. The latter ions, of course, repel one another, and so the resulting crystal is an equilibrium configuration in which the various attractions and repulsions balance out. In a NaCl crystal each Na^+ ion is surrounded by six Cl^- ions and vice versa. In crystals having different structures the number of "nearest neighbors" around each ion may be 3, 4, 6, 8, or 12. Ionic bonds are usually fairly strong, and consequently ionic crystals are strong, hard, and have high melting points.

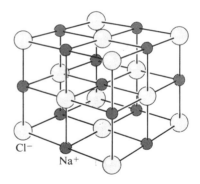

FIG. 7–5.   Ionic structure of NaCl crystal.

In covalent crystals the attractive interatomic forces arise from the sharing of electrons between atoms. Figure 7–6 shows the array of carbon atoms in a diamond crystal, with each carbon atom sharing electron pairs with the four other carbon atoms adjacent to it. All of the electrons in the outer shells of the carbon atoms participate in the binding, and it is therefore not surprising that diamond is extremely hard and

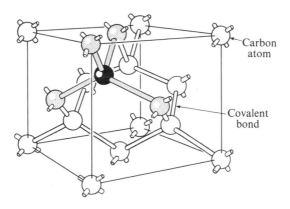

FIG. 7–6. Covalent structure of diamond. Each carbon atom shares electron pairs with four other carbon atoms. The length of each bond is 3.1×10^{-10} meter.

must be heated to over 6000°F before its crystal structure is disrupted. Other covalent crystals are those of silicon, silicon carbide, and germanium. As in the case of molecular binding, many crystalline bonds are partly ionic and partly covalent in origin. An example of such mixed binding is quartz (SiO_2).

7–3 VAN DER WAALS' BONDS

A number of molecules and nonmetallic atoms exist whose electronic structures do not lend themselves to either of the above kinds of binding. The inert gas atoms, which have filled outer electron shells, and organic molecules such as methane,

$$ H-\underset{\displaystyle\overset{\displaystyle H}{|}}{\overset{\displaystyle\overset{\displaystyle H}{|}}{C}}-H \, , $$

whose valence electrons are fully involved in the molecular bond itself, fall into this category. However, even these virtually noninteracting substances condense into solids and liquids at low enough temperatures through the action of what are known collectively as *van der Waals' forces.*

The electrostatic attraction between polar molecules that was discussed in the previous chapter is an example of a van der Waals force. These molecules have asymmetric distributions of charge, with one end positive and the other negative. When one such molecule is near another, the ends of opposite polarity attract each other to hold the molecules together (Fig. 7–7).

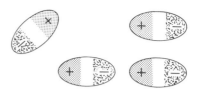

FIG. 7–7. Polar molecules attract each other.

A somewhat similar effect occurs when a polar molecule is near a nonpolar one. The electric field of the polar molecule distorts the initially symmetric charge distribution of the nonpolar one, as in Fig. 7–8, and the two then attract each other in the same way as any other pair of polar molecules. This phenomenon is similar to that involved in the attraction of bits of paper by a charged rubber comb.

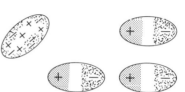

FIG. 7–8. A polar molecule attracts a non-polar one by first distorting the latter's originally symmetric charge distribution.

Curiously enough, nonpolar molecules also attract one another in the same way that polar molecules attract nonpolar ones. Because the electrons in a molecule are in constant motion, their distribution changes constantly. In a nonpolar molecule the electron distribution is symmetrical *on the average,* but *at any moment* one part of the molecule contains more electrons than usual and the remainder of the molecule contains correspondingly fewer electrons. Thus *every* molecule (and atom as well) behaves as though it is polar, although its polarization may constantly vary in both magnitude and direction. In a true polar molecule the average electron distribution is asymmetrical, while in a nonpolar molecule it is symmetrical; but in either case the molecule is polar in nature at each instant, and so is able to attract other molecules by the mechanism of Fig. 7–8. The fluctuating charge distributions of adjacent "nonpolar" molecules vary almost simultaneously so as to keep in step, as in Fig. 7–9, and the electrostatic forces that result hold the molecules

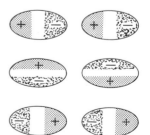

FIG. 7–9. Nonpolar molecules have, on the average, symmetric charge distributions, but at any instant the distributions are not necessarily symmetric. The fluctuations in the charge distributions of adjacent nonpolar molecules keep in step, which leads to an attractive force between them.

together provided that their random kinetic energies of thermal origin are not too great.

In general van der Waals' bonds are considerably weaker than ionic, covalent, and metallic bonds; only about 1 percent as much energy is needed to remove an atom or molecule from a van der Waals solid as is required in the case of ionic or covalent crystals. As a result the inert gases and compounds with symmetric molecules liquify and vaporize at rather low temperatures. Thus the boiling point of argon is −186°C and the boiling point of methane is −161°C. Molecular crystals, whose lattices consist of individual molecules held together by van der Waals' forces, generally lack the mechanical strength of other kinds of crystals.

7–4 THE METALLIC BOND

A fourth important type of cohesive force in crystalline solids is the *metallic bond,* which has no molecular counterpart. A characteristic property of all metal atoms is the presence of only a few electrons in their outer shells, and these electrons can be detached relatively easily to leave behind positive ions. According to the theory of the metallic bond, a metal in the solid state consists of an assembly of atoms that have given up their outermost electrons to a common "gas" of freely moving electrons that pervades the entire metal. The electrostatic interaction between the positive ions and the negative electron gas holds the metal together. This theory has many attractive features. The high electrical and thermal conductivity of metals follows from the ability of the free electrons to migrate through their crystal structures, while all of the electrons in ionic and covalent crystals are bound to particular atoms or pairs of atoms. Unlike other crystals, metals may be considerably deformed without fracture, because the electron gas permits the atoms to slide past one another in the absence of directional bonds. While certain solids, such as rubber or asphalt, that are amorphous rather than crystalline in structure can also be deformed readily, they lack the strength and hardness conferred by the metallic bond. Furthermore, since the atoms in a metal interact through the medium of a common electron gas, the properties of mixtures of different metal atoms should not depend critically on the relative proportions of each kind of atom provided their sizes are similar. This prediction is fulfilled in the observed behavior of alloys, in contrast to the specific atomic proportions characteristic of ionic and covalent solids.

The opacity and metallic luster exhibited by metals may also be traced to the gas of free electrons that pervades them. When light of any frequency shines on a metal, the free electrons are set in vibration by the oscillating electromagnetic fields and thereby absorb the light. The oscillating electrons themselves then act as sources of light, sending out electromagnetic waves of the original frequency in all directions. Those waves that happen to be directed back toward the metal surface are able to escape, and their emergence gives the metal its lustrous appearance. If the metal surface is smooth, the reradiated waves appear to us as a reflection of the original incident light.

The question of the origin of the "electron gas" in a metal is closely related to the question of why hydrogen is the only member of group I of the periodic table which is not a metal. A hydrogen molecule contains two electrons with $n = 1$ and opposite spin, the largest number possible in that state in the same atomic system without violating the Pauli exclusion principle. A hydrogen atom that approaches a H_2 molecule cannot become attached because its electron would have to go to an $n = 2$ state, which has a higher energy than the normal $n = 1$ state. The H_2 molecule exists because H_2 has less energy than $H + H$; the H_3 molecule cannot exist because H_3 has more energy than $H_2 + H$. Solid hydrogen is accordingly an assembly of H_2 molecules held together by van der Waals' bonds, bonds that are so feeble that it melts at $-259°C$.

FIG. 7–10. Each lithium atom in solid lithium has eight nearest neighbors.

Lithium, atomic number 3, is the next member of group I of the periodic table and, like all the other members except hydrogen, is indisputably metallic in its behavior. From Table 5–3 we see that each lithium atom has a filled $n = 1$ shell and a single electron in the $n = 2$, $l = 0$ state. This state can accommodate two electrons, just like the $n = 1$, $l = 0$ state of the hydrogen atom. By analogy with hydrogen, then, we expect two Li atoms to join readily to form an Li_2 molecule. However, in a Li atom there are also six vacant $n = 2$, $l = 1$ states whose energies are only a trifle higher than those of the $l = 0$ states of the same shell. A Li atom that approaches a Li_2 molecule can become attached with a covalent bond if its outermost electron goes into an $l = 1$ state, since the energy increment needed for this purpose is less than the energy liberated by the bond formation. There are a total of eight $n = 2$ states in a lithium atom, and so a lithium atom should be able to form covalent bonds with eight other lithium atoms. Lithium atoms do indeed join together in a manner in which each atom has eight nearest neighbors, as shown in Fig. 7–10.

In normal covalent substances each atom contributes one electron per bond, while in solid lithium the single outer electron of each atom must participate in eight separate bonds. To accomplish this task the electron constantly moves from bond to bond, spending only part of its time in each one; a comparable situation occurs in all metals. The electron has no way to remember from which atom it came of the pair it is between at any instant, so it is just as likely to move on to another bond that does not involve its original atom at all. The outer electrons in a lithium crystal therefore wander freely from atom to atom, constituting what is very close to the "gas" of free electrons we spoke of earlier as pervading all metals.

7–5 ATOMIC VIBRATIONS

Most solids obey Hooke's law for moderate deformations, which suggests that the atoms in a solid are held in place by interatomic bonds that also obey Hooke's law. That is, if a force F' is exerted on a particular bond to pull the atoms farther apart, the bond elongates by an amount x proportional to F':

$$F' = kx. \tag{7–2}$$

If the same force is used to push the atoms closer together, the bond contracts by the same amount x. It is possible to infer the magnitude of the interatomic force constant k in a given material from its Young's modulus Y.

We shall consider a rod L long and with a cross-sectional area A whose atoms are normally some distance a apart (Fig. 7–11). The rod has A/a^2 columns of atoms parallel to its long axis, since we can regard each atom as occupying a tiny cube a long on each edge. When a tension F is applied to the rod, then the force on each column of atoms is $F/(A/a^2)$, the total force divided by the number of columns. This force is exerted on each atomic bond in the column, and so, if we call the applied interatomic force F' as before,

$$F' = \frac{Fa^2}{A}. \tag{7–3}$$

The tension in the rod causes it to elongate by ΔL, where

$$\frac{\Delta L}{L} = \frac{1}{Y} \frac{F}{A}. \tag{7–4}$$

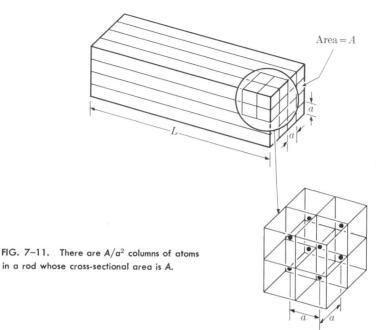

FIG. 7–11.   There are A/a^2 columns of atoms in a rod whose cross-sectional area is A.

The interatomic bonds accordingly are extended by an amount x which bears the same magnitude relative to the normal bond length a that ΔL does to the normal rod length L:

$$\frac{x}{a} = \frac{\Delta L}{L}. \tag{7–5}$$

Substituting for $\Delta L/L$ and F/A in Eq. (7–4) from Eqs. (7–5) and (7–3), respectively, we find that

$$\frac{x}{a} = \frac{1}{Y}\frac{F'}{a^2}, \qquad F' = aYx.$$

Comparing this result with Eq. (7–2) we see that the interatomic force constant k is given by

$$k = aY. \tag{7–6}$$

Interatomic distances in solids can be determined in a variety of ways. In the case of iron, for example, experiments show that

$$a = 2.5 \times 10^{-10} \text{ meter.}$$

Since Young's modulus for iron is 9.1×10^{10} n/m², the interatomic force constant in iron is

$$k = aY = 2.5 \times 10^{-10} \text{ m} \times 9.1 \times 10^{10} \text{ n/m}^2 = 23 \text{ n/m.}$$

We can calculate the frequency of the harmonic oscillations of an iron atom in a piece of solid iron by assuming that each atom behaves like a harmonic oscillator of spring constant k. This assumption is a crude one, but the behavior predicted from it is not far from what actually occurs. The mass of an iron atom is 9.3×10^{-26} kg, so that

$$f = \frac{1}{2\pi}\sqrt{\frac{k}{m}} = \frac{1}{2\pi}\sqrt{\frac{23 \text{ n/m}}{9.3 \times 10^{-26} \text{ kg}}} = 2.5 \times 10^{12} \text{ sec}^{-1}.$$

The atomic vibrations in most solids have frequencies of this order of magnitude and amplitudes at room temperature in the neighborhood of 10^{-11} meter, a small fraction of their average separation.

7–6 ENERGY BANDS

When atoms are brought as close together as those in a crystal, they interact with one another to such an extent that their outer electron shells constitute a single system of electrons common to the entire array of atoms. The Pauli exclusion principle prohibits more than two electrons (one with each spin) in any energy level of a system. This principle is obeyed in a crystal because the energy levels of the outer electron shells of the various atoms are all slightly altered by their mutual interaction. (The inner shells do not interact and therefore do not undergo a change.) As a result of the shifts in the energy levels, an *energy band* exists in a crystal in place of each sharply defined energy level of its component atoms (Fig. 7–12). While these bands are actually composed of a multi-

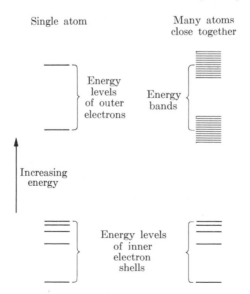

FIG. 7–12.   Energy bands replace energy levels of outer electrons in assembly of atoms close together.

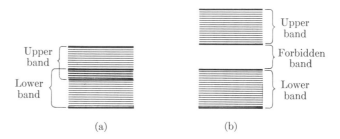

FIG. 7-13. (a) Overlapping energy bands. (b) The gap between energy bands that do not overlap is called a forbidden band.

tude of individual energy levels, as many as there are atoms in the crystal, the levels are so near one another as to form a continuous distribution.

The energy bands in a crystal correspond to energy levels in an atom, and an electron in a crystal can only have an energy that falls within one of these bands. The various energy bands in a crystal may or may not overlap, depending upon the composition of the crystal (Fig. 7-13). If they do not overlap, the gaps between them represent energy values which electrons in the crystal cannot have. The gaps are accordingly known as *forbidden bands*.

The energy bands we have been speaking of contain all of the possible energies that can be possessed by electrons. The electrical properties of a crystalline solid depend upon both its energy-band structure and the way in which the bands are normally occupied by electrons. We shall examine a few specific cases to see how the energy-band approach accounts for the observed electrical behavior of such solids.

Figure 7-14 shows the energy bands of solid sodium. Sodium atoms have but one electron in their outer shells. This means that the upper energy band in a sodium crystal is only half filled with electrons, since each level within the band, like each level in the atom, is capable of containing *two* electrons. When an electric field is established in a sodium crystal, electrons readily acquire the small additional energy they need to move up in their energy band. The additional energy is in the form of kinetic energy, and the moving electrons constitute an electric current. Sodium is therefore a good conductor, as are other crystalline solids with energy bands that are only partially filled.

Figure 7-15 shows the energy-band structure of diamond. The two lower energy bands are completely filled, and there is a gap of 6 ev between the top of the higher of these bands and the empty band above it. Hence a minimum of 6 ev of additional energy must be given to an electron in a diamond crystal if it is to be capable of free motion, since it cannot have an energy lying in the forbidden band. Such an energy increment cannot readily be imparted to an electron in a crystal by an

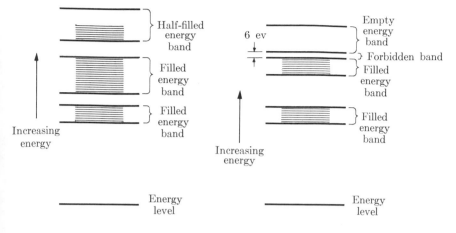

FIG. 7-14. Energy bands in solid sodium.   FIG. 7-15.   Energy bands in diamond.

electric field. A moving electron in a crystal that is not a member of a free electron gas collides with an atom or another electron after an average of only about 10^{-8} meter of travel, whereupon it loses much of the energy it gained from any electric field during its motion. The enormous electric field intensity of 6×10^8 volts/m is required if an electron is to gain 6 ev in a path length of 10^{-8} meter, well over 10^{10} times greater than the electric intensity needed to cause a current to flow in sodium. Diamond is therefore considered an electrical insulator.

Silicon has a crystal structure similar to that of diamond, and, as in diamond, a gap separates the top of a filled energy band from an empty higher band. However, while the gap is 6 ev wide in diamond, it is only 1.1 ev wide in silicon. At very low temperatures silicon is hardly better than diamond as a conductor, but at room temperature a small proportion of its electrons can possess enough kinetic energy of thermal origin to exist in the higher band. These few electrons are sufficient to permit a limited amount of current to flow when an electric field is applied. Thus silicon has a resistivity intermediate between those of conductors (such as sodium) and those of insulators (such as diamond), and is classified as a *semiconductor*.

7-7 IMPURITY SEMICONDUCTORS

The conductivity of semiconductors is markedly affected by slight amounts of impurity. Suppose we add several arsenic atoms to a silicon crystal. Arsenic atoms have five electrons in their outermost shells, while silicon atoms have four. When an arsenic atom replaces a silicon atom in a silicon crystal, four of its electrons participate in covalent bonds with its nearest neighbors. The remaining electron needs very little

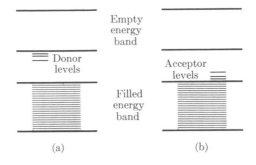

FIG. 7-16. (a) Donor levels due to arsenic atoms in a silicon crystal. (b) Acceptor levels due to gallium atoms in a silicon crystal.

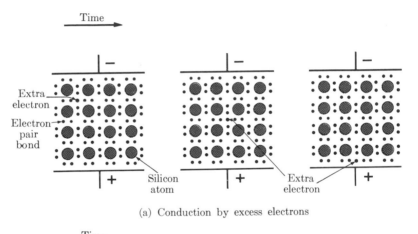

(a) Conduction by excess electrons

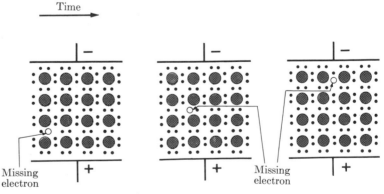

(b) Conduction by "holes"

FIG. 7-17. (a) Current in an *n*-type semiconductor is carried by excess electrons that do not fit into the electron bond structure of the crystal. (b) Current in a *p*-type semiconductor is carried by the motion of "holes," which are sites of missing electrons. Holes move toward the negative electrode as a succession of electrons move into them.

energy to be detached and move about freely in the crystal. In an
energy-band diagram, as in Fig. 7–16(a), the effect of arsenic as an
impurity is to supply energy levels just below the band in which electrons
must be present for conduction to take place. These levels are called
donor levels, and the material is called an *n-type* semiconductor because
electric current in it is carried by the motion of negative charges.

If we add gallium atoms to a silicon crystal, a different effect occurs.
Gallium atoms have only three electrons in their outer shells, and their
presence leaves vacancies called *holes* in the electron structure of the
crystal. An electron requires little energy to move into a hole, but as it
does so it leaves a new hole in its previous location. When an electric
field is applied to a silicon crystal in which gallium is present as an impurity, electrons move toward the positive electrode by successively
filling holes. The flow of current here is best described in terms of the
motion of the holes, which behave as though they are positive charges
since they move toward the negative electrode (Fig. 7–17). A material
of this kind is therefore called a *p-type* semiconductor. In the energy
band diagram of Fig. 7–16(b), we see that the effect of gallium as an
impurity is to provide energy levels, called *acceptor levels,* just above
the highest filled band; electrons that enter these levels leave behind un-
occupied levels in the formerly filled band which make possible the conduction of current.

The significance of semiconductors in technology arises from the degree
of control of electric current that can be accomplished by combining
n- and *p*-type semiconductors in various ways.

7–8 SEMICONDUCTOR DEVICES

The significance of semiconductors in technology arises from the degree
of control of electric current that can be accomplished by combining *n*-
and *p*-type semiconductors in various ways. Most semiconductor devices
depend for their action on the properties of junctions between *n*- and
p-type materials. Perhaps the most important of these properties is that
ordinarily current can flow in only one direction through a *p–n*-junction.
Figure 7–18 shows a crystal so manufactured that part exhibits *n*-type
conductivity (current carried by electrons) and the rest exhibits *p*-type
conductivity (current carried by holes). When a potential difference is
applied across the crystal so that the *p*-end is negative and the *n*-end positive, the holes in the *p*-region migrate to the left and the electrons in the
n-region migrate to the right. Only a limited number of holes and electrons are in the respective regions and new ones appear spontaneously at
only a very slow rate; the current through the entire crystal is therefore
negligible. This situation is called *reverse bias.*

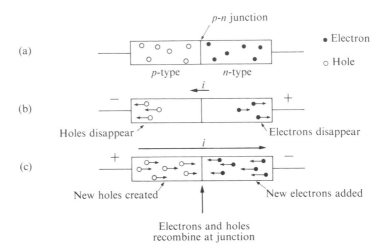

FIG. 7–18. (a) Current is carried in a p-type semiconductor by the motion of holes, and in an n-type semiconductor by the motion of electrons. (b) Reverse bias. (c) Forward bias.

Figure 7–18(c) shows the same crystal with the connections changed so that the p-end is positive and the n-end negative. Now new holes are created continuously by the removal of electrons at the p-end while new electrons are fed into the n-end of the crystal. The holes migrate to the right and the electrons to the left to produce a net positive current flowing from $+$ to $-$. The electrons and holes meet at the junction between the p- and n-regions and recombine there: a hole is a missing electron, and when electrons and holes come together they disappear into the regular structure of the crystal and can no longer act as current carriers. Thus current can flow readily through a semiconductor junction from the p- to the n-region, but hardly at all in the opposite direction. Semiconductor rectifiers are widely used today.

Transistors are semiconductor devices that, like vacuum tubes, are able to amplify weak signals into strong ones. Figure 7–19 shows an n–p–n-junction transistor, which consists of a p-type material sandwiched between n-type materials. The p-type region is called the *base*, and the two n-type regions are the *emitter* and the *collector*. In the figure the transistor is connected as a simple amplifier. There is a forward bias across the emitter-base junction in this circuit, so electrons pass readily from the emitter to the base. Depending upon the rate at which holes are produced in the base, which in turn depends upon its potential relative to the emitter and hence upon the signal input, a certain proportion of the electrons from the emitter will recombine there. The rest of the electrons migrate across the base to the collector to complete the emitter-collector circuit. If the base is at a high positive potential relative to the emitter, many holes are produced there to recombine with electrons from the

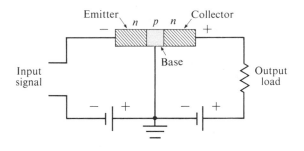

FIG. 7–19. A simple transistor amplifier.

emitter, and little current can flow through the transistor. If the relative positive potential of the base is low, the number of electrons arriving from the emitter will exceed the number of holes being formed in the base, and the surplus electrons will continue across the base to the collector. A modest change in the emitter-base voltage can cause a large change in the emitter-collector current.

The operation of a p–n–p-transistor is similar to that of an n–p–n-transistor except that holes rather than electrons are the principal current carriers.

Transistors have many advantages over vacuum tubes, notably low power consumption, high efficiency, minute size, mechanical ruggedness, and indefinite life. Transistors have already replaced vacuum tubes in a wide variety of electronic applications. However, they do have certain limitations, and cannot as yet be satisfactorily operated at very high temperatures, voltages, or currents.

7–9 FERROMAGNETISM

As a consequence of their spin, electrons behave as though they are tiny bar magnets. A quantity called *magnetic moment* is a measure of the strength of a bar magnet, so we can say that every electron possesses a certain magnetic moment. As we know, the electrons in an atom are arranged in shells about a central nucleus. A complete shell always contains an even number of electrons, half of which are oriented so that their spins point in one direction while the spins of the other half are in the opposite direction. Hence a complete shell has no net magnetic moment. In the case of all but five elements, inner and outer shells consist of electrons paired off with opposite spins except for a single electron in those shells having an odd number of electrons. In these five elements—iron, cobalt, nickel, gadolinium, and dysprosium—there are vacancies in an inner electron shell, and the electrons there are not paired off with antiparallel spins. In iron, as mentioned in the previous chapter, five of the six electrons in the $n = 3$, $l = 2$ subshell have parallel spins, so that iron atoms have appreciable magnetic moments.

A characteristic property of ferromagnetic materials is that the unpaired electrons in one atom interact strongly with their counterparts in adjacent atoms, causing all of their magnetic moments to be aligned. Atoms in these materials group themselves together in assemblies called *domains*, each about 5×10^{-5} meter across. In an unmagnetized sample, the directions of magnetization of the domains are randomly oriented, though within each domain all of the atomic moments are parallel. When such a sample is placed in an external magnetic field, either the atomic moments within the domains line up with the field or, in pure and homogeneous materials, the domain walls change so that those domains lined up with the field grow at the expense of the others. The former process requires stronger fields to take place than the latter. Hence good "permanent" magnets are inhomogeneous, for instance made from finely powdered metals sintered together, and, once magnetized, cannot change their magnetization by domain wall motion and therefore retain their magnetization. Beyond a certain temperature the atomic alignment within the domains disappears, and a ferromagnetic material becomes merely paramagnetic. This temperature, the *Curie temperature*, is 770°C in iron.

7–10 SPECIFIC HEAT

While the specific heat of a substance is a familiar notion and is generally not difficult to determine experimentally, we must use quantum theory to understand why specific heats have their observed values. This was apparent in Chapter 6, where we saw the classical kinetic-molecular theory of gases fail except in the case of a monatomic gas. A similar situation arises in the specific heats of solids.

As before, we shall consider the molar specific heat at constant volume, C_V, which is the heat that must be added to one kmole of a substance, whose volume is held constant, in order to raise its temperature by 1°C. (The specific heat at constant pressure C_P is 3 to 5 percent higher than C_V for solids.) Each of the N_0 atoms in a kmole of a solid can vibrate about its equilibrium position, and, by the principle of equipartition of energy, these vibrations in any direction represent an average of $\frac{1}{2}kT$ of kinetic and $\frac{1}{2}kT$ of potential energy for a total of kT. Since there are three possible perpendicular directions of vibration, the total thermal energy of each molecule is $3kT$. The total internal energy U of a kmole of any solid is therefore

$$U = 3N_0kT = 3RT,$$

where R is the universal gas constant. The specific heat corresponding to this expression for U is

$$C_V = \frac{dU}{dT} = 3R. \tag{7-7}$$

The value of R is 1.99 kcal/kmole·°K, and so we expect that

$$C_V = 5.97 \text{ kcal/kmole·°K.}$$

This result is called the *Dulong-Petit law*. At high temperatures the specific heats of all solids are close to the Dulong-Petit value of $3R$, and even at room temperature the agreement is not bad. At low temperatures, however, the specific heat drops sharply and tends to $C_V = 0$ at absolute zero. Figure 7–20 shows how C_V varies with absolute temperature for gold.

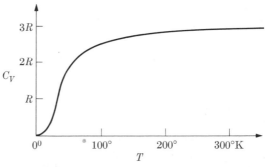

FIG. 7–20. The molar specific heat of gold approaches the Dulong-Petit value of $C_V - 3R$ at high temperatures.

The anomalous behavior of the specific heats of solids was examined by Einstein in 1907, who treated it by analogy with blackbody radiation. Einstein regarded a kmole of a solid as consisting of $3N_0$ harmonic oscillators with the natural frequency f whose energy content is quantized in multiples of hf. The average energy $\overline{E}$ of each oscillator is given by the same formula

$$\overline{E} = \frac{hf}{e^{hf/kT} - 1} \tag{7-8}$$

as that for the average energy per mode of vibration in a cavity (Section 3–2). Hence the total energy of the solid is

$$U = 3N_0\overline{E} = \frac{3N_0 hf}{e^{hf/kT} - 1}, \tag{7-9}$$

and, since $N_0 k = R$, its molar heat capacity is

$$C_V = \frac{dU}{dT} = 3R \left(\frac{hf}{kT}\right)^2 \frac{e^{hf/kT}}{(e^{hf/kT} - 1)^2}. \tag{7-10}$$

At high temperatures, when $hf \ll kT$, Eq. (7–10) becomes just $C_V = 3R$, the Dulong-Petit law. At low temperatures, when $hf \gg kT$, $e^{hf/kT}$ is much greater than 1, and

$$C_V = 3R \left(\frac{hf}{kT} \right) e^{-hf/kT}. \tag{7–11}$$

Thus Einstein's rather simple model of a solid gives a formula for C_V that approaches the Dulong-Petit value of $3R$ at high temperatures and approaches zero at low temperatures. The natural frequency f for a particular solid can be evaluated by fitting Eq. (7–10) to its empirical curve of C_V versus T (Fig. 7–10 is such a curve); a typical frequency is 5×10^{12} cycles/sec, which is reasonable.

Unfortunately the Einstein theory does not agree in detail with the exact manner in which C_V goes to zero as T goes to zero for most solids. A good fit with the experimental data requires that C_V be proportional to T^3 near $T = 0$, while Einstein's approach predicts an exponential approach as in Eq. (7–11). In 1912 Peter Debye devised a more satisfactory theory of the specific heats of solids by thinking in terms of elastic waves within the volume of a solid, which is a way to take into account the coupling between adjacent vibrating atoms. (In Einstein's model each atom was considered as independent of its neighbors.) These waves, like light waves in a cavity, are quantized in their energy content. A quantum of vibrational energy in a solid is called a *phonon*, and it travels with the speed of sound since sound is an elastic wave. ·Debye regarded a solid as a volume containing phonons of elastic energy which comprise the internal energy of the solid. When the theory of the specific heat of a solid is developed from this hypothesis, again by analogy with the theory of blackbody radiation, a result is obtained in good agreement with experiment.

The Debye theory is as successful for metals as for nonmetals, which is odd because it completely ignores the presence of free electrons in metals. In a typical metal each atom contributes one electron to the common electron gas, and in one kmole of that metal there are therefore N_0 free electrons. If these electrons behave like the molecules in an ideal gas, each would have an average of $\frac{3}{2}kT$ of kinetic energy and the metal would have

$$U_e = \tfrac{3}{2} N_0 kT = \tfrac{3}{2} RT$$

of electron energy per kmole. Hence the molar specific heat of the electrons alone would be

$$C_{Ve} = \frac{dU_e}{dT} = \frac{3}{2} R,$$

and the total molar specific heat of a metal should be (at high temperatures where the Dulong-Petit law holds)

$$C_V = 3R + \tfrac{3}{2}R = \tfrac{9}{2}R.$$

In reality $C_V = 3R$ at high temperatures, and so the free electrons do not in fact contribute to the specific heat of a metal. The problem is, why don't they?

The point was made earlier in this chapter that the electrons in a metal sample constitute a single system common to the entire sample. Electrons obey the exclusion principle, and only two of them (one of each spin) can occupy each quantum state. Thus the electron gas in a metal is not comparable with an ideal gas, where no quantum restrictions are present. At absolute zero all of the molecules in an ideal gas have zero energy, but in an electron gas the individual electrons occupy states of increasing energy. If N free electrons are present in a particular metal sample, the N lowest energy states will be filled and all those of higher energy will be empty at $T = 0°$K. The uppermost occupied energy level is called the *Fermi level* E_F. In copper, a typical metal, the Fermi level is 7.04 ev, so that in a copper sample at absolute zero electrons with energies of up to 7.04 ev will be present. The average electron energy in the sample will be 4.2 ev; if the free electrons in copper behaved classically, a temperature of nearly 50,000°K would be necessary for their average energy to be this high!

Let us now imagine the copper sample heated above $T = 0°$K. Only electrons with energies very close to E_F can acquire thermal energy, since electrons with less energy must "jump" past the higher filled quantum states in order to reach a vacant one and to do this they require much more than kT of energy. The Fermi level in copper corresponds to a temperature of 82,000°K, so an electron in the middle of the energy distribution has essentially no chance at all of ever absorbing any thermal energy at actual temperatures below the vaporization temperature of copper. At room temperature kT is about 0.02 ev, and only the very few electrons whose energies are this close to the Fermi level have much chance of absorbing thermal energy and thereby contributing to the specific heat of copper. Figure 7–21 shows the distribution of electron energies in a typical metal at $T = 0°$K, 300°K, and 1000°K; the effect of increasing the temperature is merely to round off the distribution slightly.

Another puzzling aspect of metallic behavior is answered by the quantum theory of the electron gas. The phenomenon of paramagnetism consists of the alignment of atomic magnetic moments to enhance an external magnetic field. Paramagnetism is a small effect, which is surprising in the case of metals where we would think that the free electrons could all flip over so as to align their spins with the magnetic field. However, all the available quantum states are occupied, and a given electron

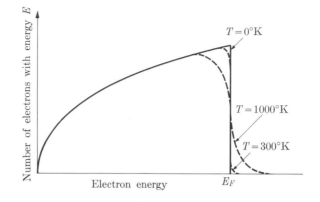

FIG. 7–21.   Distribution of electron energies in a metal at three temperatures.

cannot change its spin direction because another electron in the same energy state already has the opposite spin. Of course, given enough energy to surmount the Fermi level, the electron could align its spin with the field, but we have seen that too much energy is required for more than a minute proportion of the electrons to be able to do this.

PROBLEMS

1. (a) State the four principal binding mechanisms in solids. (b) Give an example of each of them. (c) What is the fundamental interaction responsible for each mechanism?

2. Ionic and covalent crystals are supposed to have stronger interatomic bonds than metal crystals, but everyday experience indicates that metals are, as a class, stronger than most other substances. What reason can you think of that would explain why ionic and covalent crystals cannot, as a rule, develop their theoretical strength while metals can?

3. What kind of bond holds water molecules together to form ice?

4. Since van der Waals' forces can hold inert gas atoms together to form solids, why can they not hold such atoms together to form molecules in the gaseous state?

5. What carries an electric current in a p-type semiconductor?

6. What carries an electric current in an n-type semiconductor?

7. The forbidden energy band in germanium that lies between the highest filled band and the empty band above it has a width of 0.7 ev. Compare the conductivity of germanium with that of silicon at (a) very low temperatures and (b) room temperature.

8. Does the addition of a small amount of indium to germanium result in the formation of an n- or a p-type semiconductor?

9. What is the upper limit to the contribution the free electrons in a metal can make toward its molar specific heat?

10. Because nearly all energy levels up to the Fermi level are filled in a metal, the free electrons contain a great deal of energy. Can this energy be extracted from the metal?

11. The Fermi level in silver is 5.51 ev. (a) What temperature is necessary for the average molecular energy in an ideal gas to be 5.51 ev? (b) What is the speed of a 5.51-ev electron?

8

The Nucleus

"It was shown by Bothe and Becker that some light elements when bombarded by alpha particles of polonium emit radiations which appear to be of the gamma ray type. The element beryllium gave a particularly marked effect of this kind, and later observations by Bothe, by Mme. Curie-Joliot and by Webster showed that the radiation excited in beryllium possessed a penetrating power distinctly greater than that of any gamma radiation yet found from the radioactive elements . . . Quite recently, Mme. Curie-Joliot and M. Joliot made the very striking observation that these radiations from beryllium and from boron were able to eject protons with considerable velocities from matter containing hydrogen. In their experiments the radiation from beryllium was passed through a thin window into an ionisation vessel containing air at room pressure. When paraffin wax, or other matter containing hydrogen, was placed in front of the window, the ionisation in the vessel was increased, in some cases as much as doubled. The effect appeared to be due to the ejection of protons, and from further experiment they showed that the protons had ranges in air up to about 26 cm, corresponding to a velocity of nearly 3×10^9 cm per second. They suggested that energy was transferred from the beryllium radiation to the proton by a process similar to the Compton effect with electrons, and they estimated that the beryllium radiation has a quantum energy of about 50×10^6 electron volts . . . it is difficult to account for the production of a quantum of 50×10^6 electron volts from the interaction of a beryllium nucleus and an alpha particle of kinetic energy of 5×10^6 electron volts . . . Accordingly, I made further experiments to examine the properties of the radiation excited in beryllium. It was found that the radiation ejects particles not only from hydrogen but from all other light elements which were examined. The experimental results were very difficult to explain on the hypothesis that the beryllium radiation was a quantum radiation, but followed immediately if it were supposed that the radiation consisted of particles of mass nearly equal to that of a proton and with no net charge."

James Chadwick (1891–   )

Until now we have not had to regard the nucleus of an atom as anything but a tiny positively charged lump whose sole function is to provide the atom with most of its mass and to hold its several electrons in place. Since the behavior of atomic electrons is responsible for the behavior of matter in bulk, the properties of matter we have been exploring, save for mass, have nothing directly to do with atomic nuclei. However, for all its seeming passivity, the nucleus turns out to be of supreme importance in the universe: the chemical elements exist by virtue of the ability of nuclei to possess multiple electric charges, and the energy that is involved in nearly all natural processes has its ultimate origin in nuclear reactions and transformations. Only in recent years have tools been available for investigating nuclei in any detail, and their many mysteries are still being plumbed.

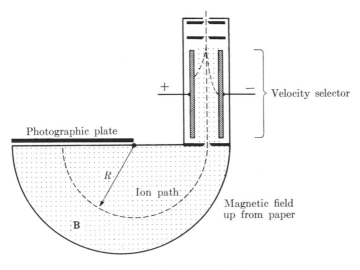

FIG. 8–1. A mass spectrometer.

8–1 THE MASS SPECTROMETER

The mass of an atom is one of its most characteristic properties and, as we shall see, an accurate knowledge of atomic masses provides considerable insight into nuclear phenomena. A variety of instruments with the generic name of *mass spectrometers* have been devised to measure atomic masses, and we shall consider the operating principles of a particularly simple one. Figure 8–1 is a schematic diagram of this mass spectrometer. The first step in its operation is to produce ions of the substance under study. If the substance is a gas, ions can be readily formed by electron bombardment, while if it is a solid it is often convenient to incorporate it into an electrode that is used as one terminal of an electric arc discharge. The ions emerge from their source through a slit with the charge $+e$ and are then accelerated by an electric field. (Ions with other charges are sometimes present but are easily taken into account.) When they enter the spectrometer itself the ions as a rule are traveling in slightly different directions with slightly different speeds. A pair of slits serves to collimate the ion beam, which then passes through a *velocity selector*. The velocity selector consists of uniform electric and magnetic fields that are perpendicular to each other and to the beam.

The electric field exerts the force

$$F_e = eE$$

on the ions, while the magnetic field exerts the force

$$F_m = Bev$$

on them in the opposite direction. For an ion to reach the slit at the far end of the velocity selector it must suffer no deflection within the selector, which means that only those ions escape for which

$$F_e = F_m ,$$

$$v = \frac{E}{B}. \tag{8-1}$$

The ions in the beam are now all moving in the same direction with the same velocity. Once past the velocity selector they enter a uniform magnetic field, and follow circular paths whose radii R may be found by equating the magnetic force Bev on them with the centripetal force mv^2/R:

$$Bev = \frac{mv^2}{R},$$

$$R = \frac{mv}{Be}. \tag{8-2}$$

Since v, B, and e are known, a measurement of R yields a value for m, the ion mass. In some spectrometer designs the ions fall upon a photographic plate, permitting R to be determined from the position of their image, while in others B is varied to bring the ion beam to a fixed detector for which R is known.

8-2 NUCLEONS

What are nuclei composed of? Let us examine a table of the properties of the elements, such as the one in Appendix D. In this table the elements are listed in the order of their *atomic numbers*, with the atomic number of an element defined as the number of electrons in each of its atoms. Hydrogen, whose atoms have a single electron each, has the atomic number 1; helium, whose atoms have two electrons each, has the atomic number 2; and so on up to lawrencium, whose atoms have 103 electrons each and which therefore has the atomic number 103. We notice immediately that atomic mass increases with atomic number, which suggests the hypothesis that all atoms are simply combinations of hydrogen atoms. According to this hypothesis (whose basic idea was put forward about 150 years ago) the helium atom, atomic number 2, should have a nucleus consisting of two protons; the lithium atom, atomic number 3, should have a nucleus consisting of three protons; and so on through the lawrencium atom, atomic number 103, which should have a nucleus consisting of 103 protons. However, atomic masses do *not* increase in steps of one hydrogen atom mass. Helium atoms weigh about four times as much as hydrogen atoms, lithium atoms about seven times as much, and lawrencium atoms about 257 times as much. So this hypothesis cannot be correct. But it must have some kind of truth in it, because, with few exceptions, atomic masses *are* very close to being exact multiples of the mass of the hydrogen atom.

What we might then suppose is that there are enough protons in each nucleus to provide for the observed atomic mass, with several electrons also present whose negative charges cancel out the positive charges. of the "excess" protons. This notion, which had favor for a while, is untenable on several grounds. We discussed one of the objections in Chapter 3: far too much energy is required to localize an electron within a nucleus, according to the uncertainty principle, for it to be a possible nuclear constituent.

In 1932 the puzzle of the missing nuclear ingredient was solved by Chadwick in England and Joliot and Curie in France. They discovered a very penetrating radiation, which was definitely not electromagnetic in character as are x-rays and gamma rays, and which readily ejected protons from hydrogen-rich materials like paraffin. The great penetrability suggested a neutral particle, which would not be affected by the strong electric fields in matter. The mass of the particle was revealed by the ease with which it could transfer energy to protons in collisions; efficient energy transfer in collisions can take place only when the participants have identical masses or very nearly so. This new particle, called the *neutron,* not only could account for the observations of Chadwick, Joliot, and Curie, but also fits perfectly into the nucleus as the uncharged but massive component required to supplement the proton. Present values for the proton and neutron masses are

$$m_{\text{proton}} = 1.6724 \times 10^{-27} \text{ kg}, \quad m_{\text{neutron}} = 1.6747 \times 10^{-27} \text{ kg}.$$

Neutrons and protons are jointly called *nucleons.*

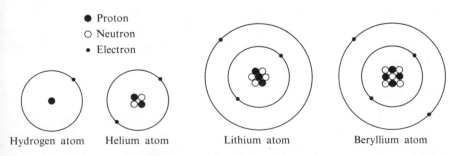

FIG. 8–2. The electronic and nuclear compositions of hydrogen, helium, lithium, and beryllium atoms.

With the help of the neutron, then, the constitution of nuclei more complex than hydrogen can be figured out readily: helium nuclei contain two protons and two neutrons each, lithium nuclei contain three protons and four neutrons each, and so on (Fig. 8–2).

8–3 ISOTOPES

We are still faced with the question of how to account for atomic masses, which cannot be exactly expressed in terms of combinations of proton and neutron masses. As it happens, the answer was known even before the composition of nuclei was clearly understood. Experimental measurements showed that what we call chemical elements actually each consist of several different components, called *isotopes*, each having the same nuclear charge (and hence the same atomic number) but with different masses. Chlorine, for example, consists of two isotopes, one whose nucleus contains 17 protons and 18 neutrons and another whose nucleus contains 17 protons and 20 neutrons. In nature, chlorine is composed of about 75% of the former isotope and 25% of the latter, yielding a composite atomic mass of 35.45, the value given in Appendix D.

Because the chemical properties of an element depend upon the distribution of the electrons in its atoms, which in turn depends upon the nuclear charge, nuclear structure beyond the number of protons present has little significance for the chemist. The physical properties of an element, however, depend strongly on the nuclear structures of its isotopes, whose behavior may be very different from one another although chemically they are indistinguishable.

The conventional symbols for isotopes follow the pattern

$$_n X^m,$$

where X is the chemical symbol of the element, n is its atomic number (equal to the number of electrons per atom or, what is equivalent, the number of protons per nucleus), and m is the *mass number* of the isotope in question, which is defined as the total number of nucleons (neutrons plus protons) in each nucleus. Hence ordinary hydrogen is designated $_1H^1$, since its atomic number and mass number are both 1, while ordinary helium is designated $_2He^4$, since its atomic number is 2 and its mass number is 4. The two isotopes of chlorine mentioned above are designated $_{17}Cl^{35}$ and $_{17}Cl^{37}$ respectively.

8–4 NUCLEAR DIMENSIONS

The Rutherford scattering experiment provides information on nuclear dimensions as well as on atomic structure. In that experiment, as we learned in Chapter 4, alpha particles are directed at a thin metal foil and their deviations measured. The observed distribution of scattering angle is consistent with a nucleus of infinitesimal size provided that the alpha particles are not too energetic. That is, below a certain particle energy the dimensions of the nucleus are small compared with the minimum distance to which the incident alpha particles approach it. At higher energies discrepancies occur between theory and data which suggest that the particles have come so close to the nucleus that it no longer can be represented as a point charge, and from the energy at which these discrepancies appear an estimate can be made of nuclear sizes.

Let us calculate the distance of closest approach r_0 of the alpha particles in the early scattering experiments. An alpha particle will come closest to a nucleus when it is headed directly at it. At the distance of closest approach the initial kinetic energy KE of the particle is entirely converted to electrostatic potential energy, and so, since the charge of the alpha particle is $2e$ and that of the nucleus is Ze (Z is the atomic number of the nucleus),

$$\text{KE} = \frac{1}{4\pi\epsilon_0}\frac{2Ze^2}{r_0}. \tag{8-3}$$

The most energetic alpha particles of natural origin have KE = 7.7 Mev, which is

$$\text{KE} = 7.7 \times 10^6 \text{ ev} \times 1.6 \times 10^{-19} \text{ j/ev} = 1.2 \times 10^{-12} \text{ j}.$$

Since $(1/4\pi\epsilon_0) = 9 \times 10^9$ n·m/coul^2, we have from Eq. (8-3),

$$
\begin{aligned}
r_0 &= \frac{1}{4\pi\epsilon_0}\frac{2Ze^2}{\text{KE}} \\
&= \frac{9 \times 10^9 \text{ n·m/coul}^2 \times 2Z \times (1.6 \times 10^{-19} \text{ coul})^2}{1.2 \times 10^{-12} \text{ j}} \\
&= 3.8 \times 10^{-16} Z \text{ m}.
\end{aligned}
$$

A typical foil material was gold, $Z = 79$. The distance of closest approach of 7.7-Mev alpha particles to gold nuclei is therefore

$$r_0 = 79 \times 3.8 \times 10^{-16} \text{ m} = 3.0 \times 10^{-14} \text{ m}.$$

The scattering of alpha particles by gold foil follows the predictions of Coulomb's law when gold nuclei are considered as point charges, from which we conclude that the radius of the gold nucleus is less than 3.0×10^{-14} meter—less than 1/10,000 of the radius of the entire atom.

When higher-energy particles are directed at thin foils, the distribution of scattering angles eventually disagrees with the theoretical expectations. The results show that the volume of a particular nucleus is proportional to its mass number A, the number of nucleons it contains. The volume of a sphere of radius R is $\frac{4}{3}\pi R^3$, so that if we consider a nucleus as spherical in shape, its radius R is proportional to $A^{1/3}$. We can write this relationship in the form

$$R = R_0 A^{1/3}, \qquad (8\text{–}4)$$

where the experimental evidence yields a value of R_0 of approximately

$$R_0 = 1.2 \times 10^{-15} \text{ m.}$$

The only isotope of gold found in nature has a mass number of 197, which means that the radius of the gold nucleus is

$$R = (197)^{1/3} \times 1.2 \times 10^{-15} \text{ m} = 5.82 \times 1.2 \times 10^{-15} \text{ m}$$
$$= 7.0 \times 10^{-15} \text{ m.}$$

Different experiments give different figures for R_0, which is why the above value was called approximate. These differences are not primarily due to experimental uncertainties but to the different ways in which a nucleus can interact with an incident particle. A neutron is acted upon exclusively by nuclear forces, for instance, while an electron is acted upon exclusively by electrical forces; protons and alpha particles are acted upon by both kinds of forces. Electron scattering leads to a smaller value for R_0 than neutron scattering, which suggests that nuclear matter and nuclear charge are differently distributed in a nucleus.

The density of nuclear matter is about $2 \times 10^{17} \text{ kg/m}^3$, a billion tons per cubic inch. "White dwarf" stars consist of atoms whose electron structures have collapsed because of immense pressures, and the densities of these stars approach the density of nuclear matter.

8–5 NUCLEAR BINDING ENERGY

While the discovery of isotopes permits us to understand why certain elements do not have atomic masses that are multiples of the proton and neutron masses, even those elements that seem to fit the picture well have slight mass variations from simple assemblies of these particles. Table 8–1 is a list of the experimentally determined masses of the neutron and some light isotopes expressed in terms of *atomic mass units* (abbreviated amu), equal to one-twelfth the mass of the $_6C^{12}$ isotope. The value of an amu is

$$1 \text{ amu} = 1.66 \times 10^{-27} \text{ kg.}$$

The deuteron and triton are nuclei of isotopes of hydrogen with mass numbers 2 and 3 respectively, which implies that they consist of one proton plus, for the deuteron, one neutron, and for the triton, two neutrons. Nuclei of $_2He^3$, an isotope of helium, consist of two protons and one neutron each, while ordinary helium nuclei consist of two protons and two neutrons each (Fig. 8–3).

TABLE 8–1

The masses of the five lightest nuclei and the neutron

Name	Symbol	Mass, amu
Proton	$_1H^1$	1.007825
Neutron	$_0n^1$	1.008665
Deuteron	$_1H^2$	2.014102
Triton	$_1H^3$	3.016049
Helium-3	$_2He^3$	3.016030
Alpha particle	$_2He^4$	4.002604

If it is true that a deuteron consists of a neutron plus a proton, having a combined mass of 2.016490 amu, why is this combined mass *more* than the 2.014102 amu mass of the deuteron? The answer is that when nucleons combine to form a nucleus, the total energy of the system (including mass energy) remains constant, although the mass may not. Thus the difference in mass between the constituents of the deuteron and the deuteron itself may appear as additional kinetic energy of the deuteron, beyond that which the neutron and proton brought along with them. Correspondingly, this same mass deficiency must be made' up by the addition of energy if it is desired to break up a deuteron.

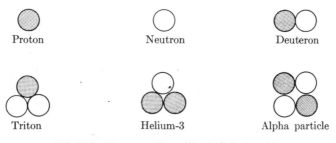

FIG. 8–3. The composition of the six lightest nuclei.

To put this concept on a quantitative basis, we recall that energy E and mass m are related by the formula

$$E = mc^2.$$

One amu is the energy equivalent of

$$1 \text{ amu} = 1.49 \times 10^{-10} \text{ j} = 9.31 \times 10^8 \text{ ev} = 931 \text{ Mev}.$$

The *binding energy* of the deuteron, which is the energy equivalent of the missing mass, is therefore

$$(2.016490 - 2.014102) \text{ amu} \times 931 \frac{\text{Mev}}{\text{amu}} = 2.22 \text{ Mev}.$$

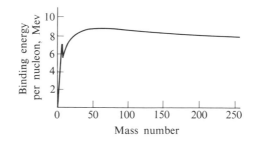

FIG. 8–4. The binding energy of the deuteron is 2.22 Mev, which means that this much energy is required to make up the difference between the deuteron's mass and the combined mass of its constituent neutron and proton. Absorbing 2.22 Mev, for instance by being struck by a 2.22-Mev gamma-ray photon, furnishes a deuteron with enough additional mass to split into a neutron and a proton.

FIG. 8–5. The binding energy per nucleon versus mass number. The higher the binding energy per nucleon, the more stable the nucleus. When a heavy nucleus is split into two lighter ones, a process called *fission*, the greater binding energy of the latter causes the liberation of energy. When two very light nuclei join to form a heavier one, a process called *fusion*, the greater binding energy of the latter again causes the liberation of energy.

This figure is confirmed by experiments that show that the minimum energy a gamma ray must have in order to disrupt a deuteron is 2.22 Mev (Fig. 8–4).

A very interesting curve results when we plot the binding energy per nucleon (that is, the total binding energy of an isotope divided by the number of neutrons and protons that compose it) versus the mass number of the nucleus (Fig. 8–5). Except for an anomalously high peak for $_2\text{He}^4$, this curve is a surprisingly regular one. The middle range of nuclei have the highest binding energy per nucleon values, about 8.75 Mev/nucleon, which means that they require the greatest amount of energy (per nucleon) to disrupt. Such nuclei are therefore the most stable.

A significant feature of nuclear structure is illustrated by this curve. Suppose that we split the nucleus $_{92}\text{U}^{235}$, whose binding energy is 7.6 Mev/nucleon, into two fragments. Each fragment will be the nucleus of a much lighter element, and therefore will have a higher binding energy per nucleon than the uranium nucleus. The difference is about 0.8 Mev/nucleon, and so, if such *nuclear fission* were to take place, an energy of

$$0.8 \; \frac{\text{Mev}}{\text{nucleon}} \times 235 \; \text{nucleons} = 190 \; \text{Mev}$$

would be given off per splitting. This is a truly immense amount of energy to be produced in a single event. As a comparison, chemical processes involve energies of the order of magnitude of one electron volt per reacting atom, 10^{-8} the energy involved in fission.

Figure 8–5 also shows that if two of the extremely light nuclei are combined to form a heavier one, the higher binding energy of the latter will also result in the evolution of energy. For instance, if two deuterons were to join to make up a $_2He^4$ nucleus, over 23 Mev would be released. This process is known as *fusion,* and, with fission, promises to be the source of more and more of the world's power.

8–6 NUCLEAR FORCES

The really surprising thing about nuclei is not that some of them are unstable, but rather that *any* of them are stable in the face of the mutually repulsive electrostatic forces exerted by the nuclear protons upon one another. There must be a nuclear force, imperceptible on a macroscopic scale, which is able to hold nuclei together. Because nuclei larger than about 10^{-15} m in radius do not exist, we may assume that the range of nuclear forces does not exceed this figure. Various experiments, in particular those involving the scattering of one nucleon by another in a collision, indicate that nuclear forces are very strong: they give rise to a binding energy between adjacent nucleons on the average over a million times greater than the 13.6 ev binding energy between the proton and electron in a hydrogen atom. Nuclear forces are evidently enormous. It is interesting that there is evidence to indicate that the proton-proton nuclear force is identical in strength and range to the neutron-proton nuclear force.

If we inquire into the "mechanism" by which two electric charges exert forces upon one another, we find that, from a formal mathematical point of view, their interaction can be described in terms of the constant circulation of electromagnetic photons between them. In 1935 the Japanese physicist Yukawa suggested that nuclear forces, too, could be regarded as the result of an interchange of certain particles between nucleons. Today these particles are called π-*mesons,* and may have charges of 0, $+e$, or $-e$. Charged π-mesons have masses of 273 times the electron mass, while neutral π-mesons have masses of 264 times the electron mass.

The crude analogy illustrated in Fig. 8–6 may help in understanding how meson exchange can lead to both attractive and repulsive forces between nuclei. Each child in the figure has a pillow. When the children exchange pillows by snatching them from each other's grasp, the effect is like that of a mutually attractive force. On the other hand, the children may also exchange pillows by throwing them at each other.

FIG. 8–6.   Particle exchange can lead to attractive or repulsive forces.

Here conservation of momentum requires that the children move apart, just as if a repulsive force were present between them.

According to Yukawa's theory, nearby nucleons constantly exchange mesons without themselves being altered. We note that the emission of a meson by a nucleon at rest which does not lose a corresponding amount of mass violates the law of conservation of energy. However, the law of conservation of energy, like all physical laws, deals only with measurable quantities. Because the uncertainty principle restricts the accuracy with which we can perform certain measurements, it limits the range of application of physical laws such as that of energy conservation. We see from the uncertainty principle in the form

$$\Delta E \ \Delta t \geq h$$

that a process can take place in which energy is not conserved by an amount ΔE provided that the time interval Δt in which the process takes place is not more than $h/\Delta E$. Thus the creation, transfer, and disappearance of a meson do not conflict with the conservation of energy provided that the sequence occurs fast enough.

The most attractive aspect of the meson theory of nuclear forces is that the π-mesons predicted by Yukawa actually exist and have the properties expected of them. For a meson to be produced which can emerge from a nucleus and be detected experimentally, enough energy must be supplied to the nucleus, usually by a collision with another particle, for conversion into meson mass without affecting the invariant masses of its nucleons. When this is done with the help of the energetic particle beams found naturally in cosmic rays or created in laboratory accelerators, mesons can be studied in detail; the results are outlined in Chapter 10. However, despite this success of the meson theory, it has not yet proved useful in understanding nuclear structure. Instead, several very different hypotheses of how nucleons behave within a nucleus seem to be required, depending upon which aspect of nuclear phenomena is involved, and we shall briefly examine two of them.

Nuclear forces, though strong, have such short ranges that each nucleon in a nucleus effectively interacts with its nearest neighbors only. This situation superficially resembles that of atoms in a solid, but the vibrations of the nucleons about their equilibrium positions would be so great as to disrupt the nucleus if its structure were like that of a crystal. A better analogy is with a liquid, whose molecules move about more or less freely while maintaining a constant intermolecular separation. The *liquid drop model* of the nucleus accounts nicely for the observed variation of nuclear binding energy with mass number A and, as we shall see in the next chapter, it provides a plausible explanation for nuclear fission.

At the opposite extreme from the picture of a nucleus as a drop of liquid is the *independent particle model*, which supposes that nucleons interact with a single force field that pervades the entire nucleus instead of only with their immediate neighbors. Thus the nucleus in this model is analogous to the atom, where only certain quantum states are permitted and, by the Pauli principle, no more than two electrons can occupy each state. Nucleons also obey the Pauli principle, so it is not altogether surprising that the independent particle model provides an understanding of some nuclear properties.

As we know, the electrons in an atom may be considered as members of various "shells," and the degree to which the outermost shell is filled governs certain aspects of the atom's behavior. Thus atoms with 2, 10, 18, 36, 54, and 86 electrons possess filled electron shells, which are stable structures. These atoms, the inert gases, do not interact with other atoms. A comparable effect is observed in the case of nuclei, where nuclei having 2, 8, 20, 28, 50, 82, and 126 neutrons or protons are exceptionally abundant. Their abundance suggests unusual stability: complex nuclei arose from reactions among lighter ones, and the evolution of larger and larger nuclei was retarded when each stable and hence inert nucleus was formed. Other evidence also suggests that the *magic numbers* 2, 8, 20, 28, 50, and 126 are significant in nuclear structure. The independent particle model of the nucleus is able to explain the origin of the magic numbers, a strong point in its favor. In recent years attempts have been made to combine the liquid drop and independent particle models into a single, consistent scheme, and much progress has been made in this direction.

PROBLEMS

1. A velocity selector uses a magnet to produce a flux density of 0.05 weber/m^2 and a pair of parallel metal plates 1 cm apart to produce a perpendicular electric field. What potential difference should be applied to the plates to permit singly charged ions of speed 5×10^6 m/sec to pass through the selector?

2. A mass spectrometer employs a velocity selector consisting of a magnetic field of flux density 0.04000 weber/m^2 perpendicular to an electric field of 50,000 volts/m. The same magnetic field is then used to deflect the ions.

Find the radius of curvature of singly charged $_3\text{Li}^7$ ions of mass 7.0160 amu in this spectrometer.

3. A beam of singly charged boron atoms is accelerated by a potential difference of 3000 volts and enters a magnetic field of flux density 0.2000 weber/m² perpendicular to the field. The ions are deflected through 180° and fall upon a photographic plate. (a) How far apart will the $_5\text{B}^{10}$ (10.013 amu) and $_5\text{B}^{11}$ (11.009) isotopes strike the photographic plate? (b) The boron found in nature is a mixture of the above isotopes, and has a composite atomic weight of 10.81 amu. Find the relative abundance of each isotope in natural boron.

4. State the number of neutrons and protons in each of the following nuclei: $_3\text{Li}^6$; $_6\text{C}^{13}$; $_{15}\text{P}^{31}$; $_{40}\text{Zr}^{94}$; $_{56}\text{Ba}^{137}$.

5. State the number of neutrons and protons in each of the following nuclei: $_5\text{Be}^{10}$; $_{10}\text{Ne}^{22}$; $_{16}\text{S}^{36}$; $_{38}\text{Sr}^{88}$; $_{72}\text{Hf}^{180}$.

6. The mass of $_{10}\text{Ne}^{20}$ is 19.9924 amu. What is its binding energy?

7. The mass of $_{17}\text{Cl}^{35}$ is 34.9689 amu. What is its binding energy?

8. Find the binding energy of the triton.

9. What is the minimum energy a gamma-ray photon must have if it is to split an alpha particle into (a) a triton and a proton, and (b) a helium-3 nucleus and a neutron?

10. Compare the electrostatic potential energy of two protons 6×10^{-15} m apart with the binding energy per nucleon of nuclei with $A \approx 60$. (Such nuclei are about 6×10^{-15} m in radius.)

11. The distance between the two protons in a $_2\text{He}^3$ nucleus is roughly 1.7×10^{-15} m. (a) Calculate the electrostatic potential energy of these protons. (b) Show that this energy is of the right order of magnitude to account for the difference in binding energy between $_1\text{H}^3$ and $_2\text{He}^3$. What conclusion can be drawn from this result about the dependence of nuclear forces upon electric charge?

9

Nuclear Transformations

"One may also imagine that in criminal hands radium might become very danger-
ous, and here we may ask ourselves if humanity has anything to gain by learning
the secrets of nature, if it is ripe to profit by them, or if this knowledge is not harmful.
The example of Nobel's discoveries is characteristic: powerful explosives have
permitted men to perform admirable work. They are also a terrible means of
destruction in the hands of the great criminals who lead the peoples towards war.
I am among those who think, with Nobel, that humanity will obtain more good than
evil from the new discoveries."

Pierre Curie (1859–1906)

While atomic nuclei are the most stable composite bodies known, they may nevertheless undergo transformations in which the identity of a particular nucleus changes. These changes may be spontaneous, as in radioactive decay, or induced, as in reactions between colliding nuclei. Aside from the data they provide on nuclear structure, nuclear transformations are noteworthy because of their role in the evolution of the universe: through them the primeval hydrogen of the universe coalesced into more complex elements, and they are responsible for virtually all of the energy that powers the universe. Today the laboratory of the nuclear physicist is as essential to astronomy as is the telescope.

9–1 RADIOACTIVITY

In the previous chapter we discussed only stable nuclei. At the beginning of the 20th century it became known, as the result of research by Becquerel, the Curies, and others, that some nuclei exist which spontaneously transform themselves into other nuclear species with the emission of radiation. Such nuclei are said to be *radioactive*. The observed radiation is of three types: *alpha particles*, which are the nuclei of $_2\text{He}^4$ atoms, *beta particles*, which are electrons, and *gamma rays*, which are composed of short-wavelength (and hence energetic) photons (Fig. 9–1).

To understand why alpha, beta, and gamma decays take place, let us examine Fig. 9–2, which is a plot showing the number of neutrons versus the number of protons in stable nuclei. For light, stable nuclei the num-

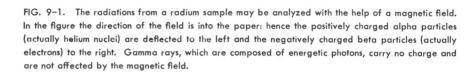

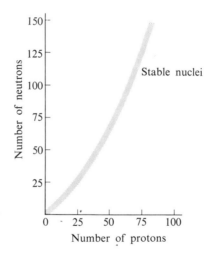

FIG. 9-1. The radiations from a radium sample may be analyzed with the help of a magnetic field. In the figure the direction of the field is into the paper: hence the positively charged alpha particles (actually helium nuclei) are deflected to the left and the negatively charged beta particles (actually electrons) to the right. Gamma rays, which are composed of energetic photons, carry no charge and are not affected by the magnetic field.

FIG. 9-2. The number of neutrons versus the number of protons in stable nuclei. The larger the nucleus, the greater the proportion of neutrons.

bers of neutrons and protons are approximately the same, while for heavier nuclei slightly more neutrons than protons are required for stability. (This is a consequence of the strong electrostatic repulsion exerted by protons on one another, which must be balanced by the attractive nuclear forces present between nucleons; to counterbalance the electrostatic repulsion of many protons, a comparatively greater number of neutrons, which produce only attractive forces owing to their electrical neutrality, is necessary for stability.) From Fig. 9-2 it is evident that an element

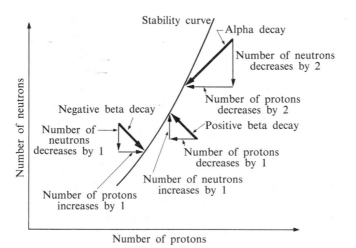

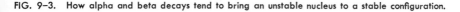

FIG. 9–3.   How alpha and beta decays tend to bring an unstable nucleus to a stable configuration.

of a given atomic number has only a very narrow range of possible numbers of neutrons if it is to be stable.

Suppose now that a nucleus exists which has either too many or too few neutrons relative to the number of protons present for stability. We shall consider the former case first. If one of the excess neutrons transforms itself into a proton, this will simultaneously reduce the number of neutrons while increasing the number of protons. To conserve electric charge, such a transformation requires the emission of a negative electron, and we may write it in equation form as

$$n^0 \rightarrow p^+ + e^-. \tag{9-1}$$

The electron leaves the nucleus, and is detectable as a "beta particle." The residual nucleus may be left with some extra energy as a consequence of its shifted binding energy, and this energy is given off in the form of gamma rays. Sometimes more than one such *beta decay* is required for a particular unstable nucleus to reach a stable configuration.

Should the nucleus have too few neutrons, the inverse reaction

$$p^+ \rightarrow n^0 + e^+, \tag{9-2}$$

in which a proton becomes a neutron with the emission of a *positive* electron (known as a *positron*), may take place. This is also called beta decay, since it resembles the emission of negative electrons from an unstable nucleus in every way save for the difference in charge.

Another way of altering its structure to achieve stability may involve a nucleus in *alpha decay*, in which an alpha particle consisting of two neutrons and two protons is emitted. Thus negative beta decay increases the number of protons by one and decreases the number of neutrons by one; positive beta decay decreases the number of protons by one and increases the number of neutrons by one; and alpha decay decreases both the number of protons and the number of neutrons by two. These processes are shown schematically in Fig. 9–3. Very often a succession of alpha and beta decays, with accompanying gamma decays to carry off excess energy, is required before a nucleus reaches stability.

Three aspects of radioactivity are wholly remarkable from the point of view of classical physics. First, the atomic number of a nucleus that undergoes alpha or beta decay changes, so that the nucleus becomes one characteristic of a different element. Elements *can* be transmuted into other elements, though hardly in a manner anticipated by alchemy. Second, radioactive decay liberates energy that can only come from *within* individual atoms. A gram of radium in a sealed container (to prevent the escape of radon, a gaseous product of its decay that is itself radioactive) evolves energy at the rate of 0.14 kcal/hr; this rate decreases so slowly that after 1600 years it has dropped by only 50 percent. Where does all of this energy come from? Not until 1905, when Einstein proposed the equivalence of mass and energy, was this puzzle understood. Third, as we shall see in the next section, radioactivity is a statistical process. Every nucleus of a radioisotope has a certain likelihood of decaying, but, because the decay obeys the laws of chance, we have no way of predicting *which* nuclei will actually decay at a particular time. There is no cause-effect relationship in radioactivity as there is in all of classical physics. The revolution in physics that finally overthrew determinism in 1925 with the advent of quantum theory thus had its beginnings several decades earlier in the discovery of radioactive decay.

9–2 HALF-LIFE

The rate at which a sample of a radioactive material decays is called its *activity*, R. If there are N undecayed nuclei in the sample at a particular time,

$$R = -\frac{dN}{dt}. \tag{9–3}$$

Because N decreases with time, dN/dt itself is negative, and the minus sign is inserted to make R a more convenient positive quantity. According to Eq. (9–3), R should be expressed in disintegrations/sec, but the activities encountered in practice are usually so high that a larger unit, the *curie*, is more often used. Originally the curie was defined as the activity of 1 gm of radium ($_{88}Ra^{226}$), and its precise value accordingly changed as measuring techniques improved. For this reason the curie is now defined arbitrarily as

$$1 \text{ curie} = 3.70 \times 10^{10} \text{ disintegrations/sec};$$

the activity of 1 gm of radium is a few percent smaller. The *millicurie* (10^{-3} curie) and *microcurie* (10^{-6} curie) are frequently employed to supplement the curie. A luminous watch dial contains several microcuries of $_{88}Ra^{226}$; ordinary potassium has an activity of about 1 millicurie/kg owing to the presence of the radioactive isotope $_{19}K^{40}$; "cobalt-60" sources of 1 or more curies are widely used in medicine for radiation therapy and industrially for the inspection of metal castings and welded joints.

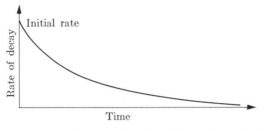

FIG. 9-4. The rate at which a sample of a radioactive substance decays is not constant, but varies with time in the manner shown in the curve. Some isotopes decay faster than others, but all obey curves having the same shape.

One of the characteristics of radioactivity is that the rate at which the nuclei in a given sample decay always follows a curve whose shape is like that shown in Fig. 9-4. Some isotopes decay faster than others, but in each case a certain definite time is required for half of an original sample to decay. This time is called the *half-life*, $T_{1/2}$, of the isotope. Observed half-lives range from billionths of a second to billions of years. Radon ($_{86}Rn^{222}$) has a half-life against alpha decay of 3.8 days, for instance: if we start with 1 gm of radon, $\frac{1}{2}$ gm will remain after 3.8 days, $\frac{1}{4}$ gm after 7.6 days, $\frac{1}{8}$ gm after 11.4 days, and so on (Fig. 9-5). This kind of decrease in activity with time is exponential. That is, if the activity of a radioactive sample at a certain moment is R_0, at a time t later it will be

$$R = R_0 e^{-\lambda t}. \qquad (9-4)$$

The quantity λ, the *decay constant*, has a unique value for each radioisotope.

Let us investigate the connection between decay constant λ and half-life $T_{1/2}$. According to the definition of half-life, when $t = T_{1/2}$, $R = \frac{1}{2}R_0$. Hence, substituting these figures into Eq. (9-4), we have

$$\tfrac{1}{2}R_0 = R_0 e^{-\lambda T_{1/2}},$$

$$e^{\lambda T_{1/2}} = 2.$$

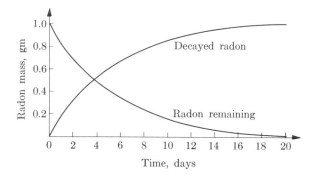

FIG. 9-5.   The decay of radon, whose half-life is 3.8 days.

Taking the natural logarithm of both sides gives

$$\lambda T_{1/2} = \ln 2 = 0.693$$

$$\lambda = \frac{0.693}{T_{1/2}}.$$ (9-5)

The decay constant of radon, which has a half-life of 3.8 days, is therefore

$$\lambda = \frac{0.693}{T_{1/2}} = \frac{0.693}{3.8 \text{ days} \times 86,400 \text{ sec/day}}$$

$$= \frac{0.693}{3.3 \times 10^5 \text{ sec}} = 2.1 \times 10^{-6} \text{ sec}^{-1}.$$

Radioactive decay involves individual events taking place within independent nuclei, rather than collective processes that involve more than one nucleus in interaction. This idea is confirmed by experiments which show that the half-life of a particular isotope is absolutely invariant under changes of pressure, temperature, electric and magnetic fields, and so on, which might, if sufficiently strong, influence internuclear phenomena. Samples of radioisotopes decay exponentially because a great many individual nuclei are involved, each having a certain probability of decaying per unit time. If λ is this probability, $\lambda \, dt$ is the probability that one of these nuclei will decay in the time interval dt. Hence the actual number dN of decays that occur in dt is the product of $\lambda \, dt$ and N, the total number of nuclei present in the sample at the start of the time interval:

$$dN = -N\lambda \, dt, \qquad \frac{dN}{N} = -\lambda \, dt.$$ (9-6)

(The minus sign is necessary because N is a decreasing function of time.) Integrating the left-hand side of Eq. (9–6) from the initial number of nuclei N_0 at $t = 0$ to the final number N after the time t has elapsed and the right-hand side from $t = 0$ to $t = t$ yields

$$\int_{N_0}^{N} \frac{dN}{N} = -\lambda \int_0^t dt,$$

$$\ln N - \ln N_0 = -\lambda t,$$

$$N = N_0 e^{-\lambda t}. \tag{9–7}$$

We can go from this formula, which gives the number of undecayed nuclei at the time t in terms of the original number N_0 and the decay probability λ, to the empirical formula of Eq. (9–4) for the activity of a radioactive sample with the help of the definition of activity. From Eq. (9–3),

$$R = -\frac{dN}{dt} = N_0 e^{-\lambda t}. \tag{9–8}$$

This result is the same as Eq. (9–4) if we identify R_0, the initial activity, with λN_0 and the decay constant λ with the decay probability per nucleus per unit time.

We can use the relationship

$$R = \lambda N \tag{9–9}$$

among activity, decay constant, and number of nuclei to calculate the activity of a given radioactive sample, for instance that of a milligram (mgm) of radon. A kmole of radon has a mass equal to the atomic mass of radon, 222, expressed in kg. One mgm of radon, which is 10^{-6} kg, therefore represents

$$\frac{10^{-6} \text{ kg}}{222 \text{ kg/kmole}} = 4.5 \times 10^{-9} \text{ kmole.}$$

Because 1 kmole of any isotope contains Avogadro's number of atoms, 1 mgm of radon contains

$$4.5 \times 10^{-9} \text{ kmole} \times 6.02 \times 10^{26} \text{ atoms/kmole} = 2.7 \times 10^{18} \text{ atoms.}$$

Hence N is 2.7×10^{18} here and, as we found earlier, λ is 2.1×10^{-6} sec^{-1} for radon. Substituting these figures in Eq. (9–9) we find that the activity of 1 mg of radon is

$$R = \lambda N = 2.1 \times 10^{-6} \text{ sec}^{-1} \times 2.7 \times 10^{18}$$

$$= 5.7 \times 10^{12} \text{ disintegrations/sec}$$

$$= 150 \text{ curies.}$$

The half-life of a radioisotope is not the same as its average lifetime $\overline{T}$ as its activity decreases exponentially and not linearly with time. The average lifetime is simply the reciprocal of the decay probability per unit time λ. In view of the relationship between λ and half-life $T_{1/2}$,

$$\overline{T} = \frac{1}{\lambda} = \frac{T_{1/2}}{0.693} = 1.44\ T_{1/2}.$$

The mean lifetime of radon, whose half-life is 3.8 days, is 5.5 days.

9–3 ALPHA DECAY

Nuclei that contain more than about 210 nucleons are so large that the short-range forces holding them together are barely able to counterbalance the long-range electrostatic repulsive forces of their protons. Such a nucleus can reduce its bulk and thereby achieve greater stability by emitting an alpha particle, which decreases its mass number A by 4. It is appropriate to ask why it is that only alpha particles are given off by excessively heavy nuclei, and not, for example, individual protons or $_2\text{He}^3$ nuclei. The reason is a consequence of the high binding energy of the alpha particle, which means that it has significantly less mass than four individual nucleons. Because of this small mass, an alpha particle can be ejected by a heavy nucleus with energy to spare. Thus the alpha particle released in the decay of $_{92}\text{U}^{232}$ has a kinetic energy of 5.4 Mev, while 6.1 Mev would have to be supplied from the outside to this nucleus if it is to release a proton, and 9.6 Mev supplied if it is to release a $_2\text{He}^3$ nucleus.

Even though alpha decay may be energetically possible in a particular nucleus, it is not obvious just how the alpha particle is able to break away from the nuclear forces that bind it to the rest of the nucleus. Typically an alpha particle has available about 5 Mev of energy with which to escape. However, an alpha particle located at a point near the nucleus but just outside the range of its nuclear forces has an electrostatic potential energy of perhaps 25 Mev; that is, if released from this position it will have a kinetic energy of 25 Mev when it is an infinite distance away as a result of electrostatic repulsion. An alpha particle inside the nucleus therefore should require a minimum of 25 Mev in energy, five times more than is available, in order to break loose (Fig. 9–6). The alpha particle, then, is located in a box whose walls are of such a height that an energy of 25 Mev is needed to surmount them, while the particle itself has only 5 Mev for the purpose.

Quantum mechanics provides the answer to the paradox of alpha decay. Two assumptions are needed: (1) an alpha particle can exist as an individual entity within a nucleus, and (2) it is in constant motion there. According to quantum theory, a moving particle has a wave

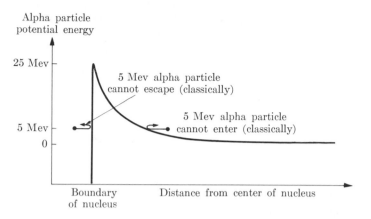

FIG. 9–6. The variation in alpha-particle potential energy near a typical heavy nucleus. The alpha particle in this nucleus has 5 Mev available for its escape.

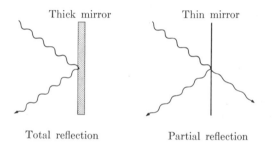

FIG. 9–7. In the process of reflection, a wave penetrates the reflecting surface for a short distance and may pass through it if the mirror is sufficiently thin.

character, so that the proper classical analog of an alpha particle in a nucleus is a light wave trapped between mirrors and not a particle bouncing back and forth between solid walls. Now in order for a light wave to be reflected from a mirror, it must actually penetrate the reflecting surface for a short distance (Fig. 9–7). The intensity of the wave drops off quite rapidly inside the reflecting surface, to be sure, but it *must* penetrate to some extent. If the mirror is thick, all of the incident light is reflected. However, if the mirror is very thin, some of the incident light can pass right through the mirror, as shown. The formal theory of this partial transmission, with only minor changes, is able to account quantitatively for alpha decay. The very existence of alpha decay, in fact, is further confirmation of the validity of quantum ideas, since the principles of physics that follow from Newton's laws of motion prohibit such decay.

Of course, a 25-Mev energy barrier is not very "transparent" to a 5-Mev alpha particle. A typical heavy nucleus might be 2×10^{-14} meters in diameter, and an alpha particle within it might oscillate back and forth with a speed of 2×10^7 m/sec. Hence the alpha particle strikes the confining nuclear wall 10^{21} times per second, but may nevertheless have to wait as much as 10^{10} years to escape from certain nuclei! The quantum-mechanical phenomenon of barrier penetration is sometimes called the *tunnel effect* because the particle escapes *through* the barrier and not over it.

9–4 NUCLEAR REACTIONS

When two nuclei come close enough together, it is possible for a rearrangement of their constituent nucleons to occur with one or more new nuclei formed. These are called *nuclear reactions,* by analogy with chemical reactions in which two or more compounds may combine to form new ones.

An immense number of nuclear reactions is known, all of which obey certain conservation laws. The three most important of these laws are: (1) conservation of electric charge; (2) conservation of number of nucleons; and (3) conservation of mass energy. The reaction

$$_3\text{Li}^7 + {}_1\text{H}^1 \rightarrow 2 \; _2\text{He}^4 \qquad\qquad (9\text{–}10)$$

involves the fusion of a lithium nucleus of atomic number 3 and mass number 7 with a proton of atomic number 1 and mass number 1 (Fig. 9–8). This reaction results in two alpha particles of atomic number 2 and mass number 4. A total of four protons is present on each side of the arrow, satisfying condition (1). Further, eight nucleons are present on each side of the arrow, taking care of condition (2). To see whether (3) is satisfied, we note that the mass of $_3\text{Li}^7$ is 7.01600, the mass of $_1\text{H}^1$

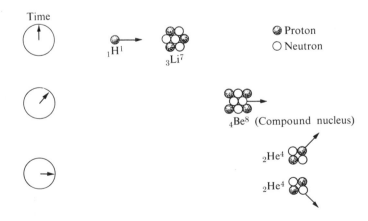

FIG. 9–8. The fusion reaction in which a lithium nucleus struck by a proton absorbs the proton and thereupon splits into two alpha particles. The liberated energy of 17.3 Mev is carried away by the alpha particles as kinetic energy.

is 1.00783, and the mass of $_2\text{He}^4$ is 4.00260, all in atomic mass units. Thus the mass excess Q of the left-hand side of Eq. (9–10) is

$$Q = 7.01600 \text{ amu} + 1.00783 \text{ amu} - 2 \times 4.00260 \text{ amu}$$

$$= 0.01863 \text{ amu}$$

$$= 17.3 \text{ Mev.}$$

This energy is divided between the two alpha particles.

It is important to note that the $_3\text{Li}^7$ and proton must have some initial relative velocity, hence initial energy, in order that they approach close enough for the reaction to take place. This initial energy must be distinguished from the initial energy required by a process whose products have *more* mass than the reactants. Let us examine the reaction

$$_7\text{N}^{14} + _2\text{He}^4 \rightarrow _8\text{O}^{17} + _1\text{H}^1. \tag{9–11}$$

Here, since the masses of the nuclei involved are respectively 14.00307, 4.00260, 16.99913, and 1.00783 amu, there is a mass *deficiency* of 0.00128 amu, which is 1.2 Mev. This energy must be supplied by the kinetic energy of either the alpha particle or the nitrogen nucleus or both if the reaction is to occur at all.

Often nuclear reactions occur in two separate phases. The first phase occurs when a particle (a nucleon or another nucleus) collides with a nucleus and joins with it to form a composite body called a *compound nucleus* (Fig. 9–8). In the second phase the compound nucleus decays into the products of the reaction. The notion of the compound nucleus is a useful one because a particular compound nucleus may be formed in a variety of ways, but it will decay in a characteristic manner that is independent of the way in which it came into being. The amount of excitation energy possessed by the compound nucleus is what determines how it will decay. The excitation energy is the energy equivalent of the mass difference between the initial particles and the compound nucleus plus whatever kinetic energy was brought in by the incident particle. For example, when $_7\text{N}^{14}$ is formed by the collision of a 6-Mev proton with a $_6\text{C}^{13}$ nucleus, its excitation energy is the sum of 6 Mev and the 7.5-Mev mass energy difference between $_7\text{N}^{14}$ and $_6\text{C}^{13} + _1\text{H}^1$. The compound nucleus $_7\text{N}^{14}$ whose excitation energy is 13.5 Mev usually decays into $_7\text{N}^{13}$ and a neutron:

$$_7\text{N}^{14} + 13.5 \text{ Mev} \rightarrow _7\text{N}^{13} + _1\text{n}^0 + 3 \text{ Mev.} \tag{9–12}$$

The 3 Mev that is liberated as kinetic energy is the difference between the 13.5-Mev excitation energy and the mass energy difference between $_7\text{N}^{14}$ and $_7\text{N}^{13} + _0\text{n}^1$. The essential point here is that the decay reaction (9–12) occurs regardless of how the $_7\text{N}^{14}$ compound nucleus acquired its 13.5 Mev of excitation; a 2-Mev alpha particle can react with a $_5\text{B}^{10}$

nucleus to give the identical compound nucleus, as can a 3.2-Mev deuteron with a $_6C^{12}$ nucleus.

The probability that a given pair of nuclear particles will react to form a compound nucleus depends both upon the identities of the particles and upon the energy brought into their interaction, even though the various conservation laws permit the reaction to take place. For example, $_{48}Cd^{113}$ is almost 10,000 times more likely to capture a 0.1-ev neutron than a 10-ev neutron, while it is highly improbable that $_{48}Cd^{112}$ will capture a neutron of either energy at all.

9-5 NUCLEAR FISSION

A particularly significant nuclear reaction that requires a neutron to initiate it is *fission*. In this process, which can take place only in certain very heavy nuclei such as $_{92}U^{235}$, the absorption of an incoming neutron causes the target nucleus to split into two smaller nuclei called *fission fragments* (Fig. 9-9). Because stable light nuclei have proportionately fewer neutrons than do heavy nuclei (Fig. 9-2), the fragments are unbalanced when they are formed and at once release one or two neutrons each. Usually the fragments are still somewhat unstable, and may undergo beta decays (accompanied by gamma decays) to achieve appropriate neutron:proton ratios. The products of fission, such as the fallout from a nuclear bomb burst, are accordingly highly radioactive. Although

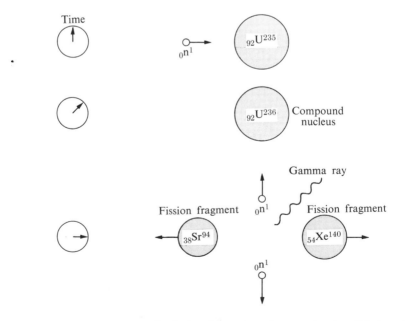

FIG. 9-9. In nuclear fission, an absorbed neutron causes a heavy nucleus to split in two parts, with the emission of several neutrons and gamma rays.

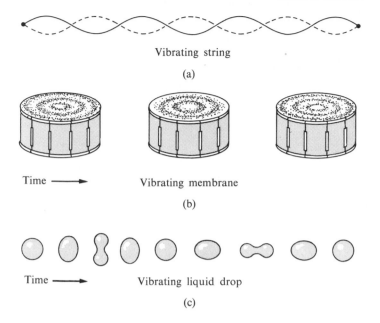

Vibrating string

(a)

Time ⟶   Vibrating membrane

(b)

Time ⟶   Vibrating liquid drop

(c)

FIG. 9–10. The characteristic oscillations of (a) a one-dimensional body, (b) a two-dimensional body, and (c) a three-dimensional body.

a variety of nuclear species may appear as fission fragments, we might cite as a typical fission reaction

$$_{92}U^{235} + _0n^1 \rightarrow _{92}U^{236} \rightarrow _{54}Xe^{140} + _{38}Sr^{94}$$
$$+ _0n^1 + _0n^1 + \gamma + 200 \text{ Mev.} \qquad (9\text{--}13)$$

The $_{92}U^{236}$ which is first formed is a compound nucleus, and it is this nucleus that splits in two. The fission fragments $_{54}Xe^{140}$ and $_{38}Sr^{94}$ are both beta radioactive; the former decays four successive times until it becomes the stable isotope $_{58}Ce^{140}$, and the latter decays twice in becoming the stable isotope $_{40}Zr^{94}$. About 84% of the total energy liberated during fission appears as kinetic energy of the fission fragments, about 2.5% as kinetic energy of the neutrons, and about 2.5% in the form of instantaneously emitted gamma rays, with the remaining 11% being given off in the decay of the fission fragments.

The liquid drop model of the nucleus that was mentioned in the previous chapter provides a plausible mechanism for the fission process. We are all familiar with the characteristic oscillations of a stretched string, Fig. 9–10(a), and of a taut membrane, Fig. 9–10(b). Less familiar perhaps are the characteristic oscillations of a liquid drop, shown in Fig. 9–10(c). If we accept the picture of a nucleus as a liquid drop, we can suppose that the absorption of a neutron by a heavy nucleus is enough to set it vibrating. The difference between an ordinary liquid drop and

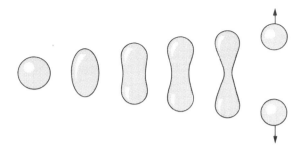

FIG. 9–11.   Fission according to the liquid drop model.

a nucleus is that, when the latter is distorted from a spherical shape, the short-range nuclear forces holding it together lose much of their effectiveness owing to the larger nuclear surface where nucleons have fewer bonds. Though the distortion may seriously weaken the attractive forces in the nucleus, the repulsive electrostatic forces of the protons are only slightly affected. If the attractive forces still predominate, the excitation energy of the nucleus eventually is lost through gamma decay: a neutron is captured and a gamma ray is emitted. If the repulsive forces predominate, however, the distortion grows larger and larger until the nucleus splits in two (Fig. 9–11), which is observed as fission.

Because each fission event liberates two or three neutrons while only one fission is required to initiate it, a rapidly multiplying sequence of fissions can occur in a lump of suitable material (Fig. 9–12). When

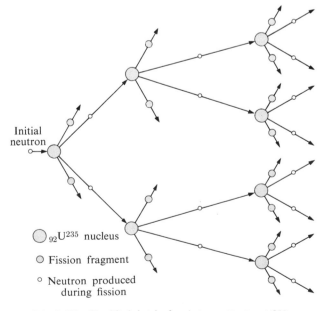

Initial
neutron

$_{92}U^{235}$ nucleus

Fission fragment

Neutron produced
during fission

FIG. 9–12.   Simplified sketch of a chain reaction in $_{92}U^{235}$.

uncontrolled, such a *chain reaction* evolves an immense amount of energy in a short time. If we assume that two neutrons emitted in each fission are able to induce further fissions (the average figure is lower in practice) and that 10^{-8} sec elapses between the emission of a neutron and its subsequent absorption, a chain reaction starting with a single fission will release 2×10^{13} joules of energy in less than 10^{-6} sec! An uncontrolled chain reaction evidently can cause an explosion of exceptional magnitude. When properly controlled so as to assure that exactly one neutron per fission causes another fission, a chain reaction occurs at a constant level of power output. A reaction of this kind makes a very efficient source of power: an output of about 1000 kilowatts is produced by the fission of 1 gram of a suitable isotope per day, as compared with the consumption of over 3 tons of coal per day per 1000 kilowatts in a conventional power plant. A device in which a chain reaction can be initiated and controlled is called a *nuclear reactor*.

9–6 NUCLEAR FUSION

Virtually all of the energy in the universe originates in the fusion of hydrogen nuclei into helium nuclei in stellar interiors, where hydrogen is the most abundant element. Two different reaction sequences are possible, with the likelihood of each depending upon the properties of the star involved. The *proton-proton cycle* is the predominant energy source of stars whose interiors are cooler than that of the sun, perhaps 2×10^6 °K. The proton-proton cycle proceeds by means of the following reactions:

$$_1\mathrm{H}^1 + {}_1\mathrm{H}^1 \rightarrow {}_1\mathrm{H}^2 + e^+ + 0.4 \,\mathrm{Mev},$$
$$_1\mathrm{H}^1 + {}_1\mathrm{H}^2 \rightarrow {}_2\mathrm{He}^3 + 5.5 \,\mathrm{Mev}, \tag{9–14}$$
$$_2\mathrm{He}^3 + {}_2\mathrm{He}^3 \rightarrow {}_2\mathrm{He}^4 + 2\,{}_1\mathrm{H}^1 + 12.9 \,\mathrm{Mev}.$$

The first two of these reactions must each occur twice for every synthesis of $_2\mathrm{He}^4$, so that the total energy produced per cycle is 24.7 Mev.

Stars hotter than the sun obtain their energy from the *carbon cycle*. This cycle requires a $_6\mathrm{C}^{12}$ nucleus for its first step and in its last step regenerates a $_6\mathrm{C}^{12}$ nucleus, so that this isotope may be thought of as a catalyst for the process. The carbon cycle proceeds as follows:

$$_1\mathrm{H}^1 + {}_6\mathrm{C}^{12} \rightarrow {}_7\mathrm{N}^{13} + 2.0 \,\mathrm{Mev},$$
$$_7\mathrm{N}^{13} \rightarrow {}_6\mathrm{C}^{13} + e^+ + 1.2 \,\mathrm{Mev},$$
$$_6\mathrm{C}^{13} + {}_1\mathrm{H}^1 \rightarrow {}_7\mathrm{N}^{14} + 7.6 \,\mathrm{Mev},$$
$$_7\mathrm{N}^{14} + {}_1\mathrm{H}^1 \rightarrow {}_8\mathrm{O}^{15} + 7.3 \,\mathrm{Mev}, \tag{9–15}$$
$$_8\mathrm{O}^{15} \rightarrow {}_7\mathrm{N}^{15} + e^+ + 1.7 \,\mathrm{Mev},$$
$$_7\mathrm{N}^{15} + {}_1\mathrm{H}^1 \rightarrow {}_6\mathrm{C}^{12} + {}_2\mathrm{He}^4 + 4.9 \,\mathrm{Mev}.$$

Here again the net result is the formation of an alpha particle and two positrons from four protons with a total of 24.7 Mev of energy evolved. In the sun both the proton-proton and carbon cycles take place with comparable probabilities.

The energy liberated by nuclear fusion is often called *thermonuclear energy*. High temperatures and densities are necessary for fusion reactions to occur in such quantity that a substantial amount of thermonuclear energy is produced: the high temperature assures that the initial light nuclei have enough thermal energy to overcome their mutual electrostatic repulsion and come close enough together to react, while the high density assures that such collisions are frequent. A further condition for the proton-proton and carbon cycles is a large reacting mass, such as that of a star, since a number of separate steps is involved in each cycle and much time may elapse between the initial fusion of a particular proton and its ultimate incorporation in an alpha particle. On the earth, where any reacting mass must be very limited in size, an efficient thermonuclear process cannot involve more than a single step. The reactions that appear most promising as sources of commercial power involve the combination of two deuterons to form a triton and a proton,

$$_1\text{H}^2 + {}_1\text{H}^2 \rightarrow {}_1\text{H}^3 + {}_1\text{H}^1 + 4.0 \text{ Mev}, \qquad (9\text{–}16)$$

or their combination to form a $_2\text{He}^3$ nucleus and a neutron,

$$_1\text{H}^2 + {}_1\text{H}^2 \rightarrow {}_2\text{He}^3 + {}_0\text{n}^1 + 3.3 \text{ Mev}. \qquad (9\text{–}17)$$

Both reactions have about equal probabilities. A major advantage of reactions (9–16) and (9–17) is that deuterium (the variety of hydrogen whose atomic nuclei are deuterons, sometimes called "heavy hydrogen") is relatively abundant on the earth, so that there should be no fuel problems in power plants operating on deuteron fusion. While there are many difficulties to surmount in achieving practical thermonuclear power, it will almost certainly become an eventual reality.

PROBLEMS

1. The half-life of radium is 1600 years. How long will it take for 15/16 of a given sample of radium to decay?

2. Sixty hours after a sample of the beta emitter $_{11}\text{Na}^{24}$ has been prepared, only 6.25% of it remains undecayed. What is the half-life of this isotope?

3. The hydrogen isotope tritium, $_1\text{H}^3$, is radioactive and emits an electron with a half-life of 12.5 years. (a) What does $_1\text{H}^3$ become after beta decay? (b) What percentage of an original sample of tritium will remain 25 years after its preparation?

4. One millicurie of $_{82}\text{Pb}^{214}$ has a mass of 3×10^{-14} kg. Find its decay constant.

5. The half-life of $_{92}U^{238}$ against alpha decay is 4.5×10^9 years. Find the number of disintegrations per second that occur in 1 kg of this isotope.

6. The isotope $_{92}U^{239}$ undergoes two negative beta decays in becoming an isotope of plutonium. State the atomic number and mass number of this isotope.

7. The nuclei $_8O^{14}$ and $_8O^{19}$ both undergo beta decay in order to become stable nuclei. Which would you expect to emit a positron and which an electron?

8. The nucleus $_6C^{11}$ undergoes positive beta decay. What is the atomic number, mass number, and chemical name of the resulting nucleus?

9. The isotope $_{90}Th^{233}$ undergoes two negative beta decays in becoming an isotope of uranium. State the atomic number and mass number of this isotope.

10. The nucleus $_{92}U^{238}$ decays into a lead isotope through the successive emissions of eight alpha particles and six electrons. State the atomic number and mass number of the lead isotope.

11. Radium undergoes spontaneous decay into helium and radon. Why is radium regarded as an element rather than as a chemical compound of helium and radon?

12. How much energy is evolved in the nuclear reaction $_1H^2 + _1H^3 \rightarrow _2He^4 + _0n^1$?

13. The nuclear reaction $_3Li^6 + _1H^2 \rightarrow 2 _2He^4$ evolves 22.4 Mev. Using the mass values of the deuteron and alpha particle given in Table 26–1, calculate the mass of $_3Li^6$ in amu.

14. In certain stars three alpha particles join in a single reaction to form a $_6C^{12}$ nucleus. The mass of $_6C^{12}$ is 12.0000 amu. How much energy is evolved in this reaction?

15. What are the differences and similarities between nuclear fusion and nuclear fission?

16. A nucleus of $_7N^{15}$ is struck by a proton. A nuclear reaction takes place with the emission of (a) a neutron, or (b) an alpha particle. Give the atomic number, mass number, and chemical name of the remaining nucleus in each of the above cases.

17. A nucleus of $_4Be^9$ is struck by an alpha particle. A nuclear reaction takes place with the emission of a neutron. Give the atomic number, mass number, and chemical name of the resulting nucleus.

18. A reaction often used to detect neutrons occurs when a neutron strikes a $_5B^{10}$ nucleus, with the subsequent emission of an alpha particle. What is the atomic number, mass number, and chemical name of the remaining nucleus?

19. If each fission in $_{92}U^{235}$ releases 200 Mev, how many fissions must occur per second to produce a power of 1 kw?

10

Elementary Particles

"Seek simplicity, and distrust it."
"Less is more."

Alfred Whitehead (1861–1947)
Ludwig Mies van der Rohe (1886–   )

The notion that the physical universe consists of myriad particles of a few basic kinds and that its evolution can be traced to the interactions of these particles originated long ago. Today, though convinced of the soundness of this notion, we are still far from a full understanding of these particles and of the reasons for their behavior. The existence of elementary particles was proposed in an effort to find order in the diversity of nature, and their generic name suggests their role as the ultimate, indivisible entities that constitute reality. Unfortunately for the quest for order and simplicity, a total of several score different "elementary" particles have been discovered, and, though relatively few of them are stable when alone in space, all contribute in some way to the structure and behavior of matter. In this chapter we shall look into the evidence for these particles and into some ideas on their possible functions.

10–1 THE NEUTRINO

In radioactive decay, as in all other natural processes, energy (including mass energy) is conserved. For this reason the total mass of the products of a particular decay must be less than the mass of the initial nucleus, with the missing mass appearing as photon energy in the case of gamma decay and as kinetic energy in the cases of alpha and beta decay. In gamma and alpha decay the liberated energy is indeed precisely equal to the energy equivalent of the lost mass, but in beta decay a strange effect occurs: instead of all having the same energy, the emitted electrons from a particular isotope exhibit a variety of energies. These energies range from zero up to a maximum figure equal to the energy equivalent of the missing mass in the transformation. This effect is illustrated in Fig. 10–1, which shows the spread in electron energy the decay of $_{83}Bi^{210}$.

Momentum as well as energy is apparently not conserved in beta decay. When an object at rest disintegrates into two parts, they must move apart in opposite directions in order that the total momentum of the system remain zero. Experiments show, however, that the emitted electron and the residual nucleus do *not* in general travel in opposite directions after beta decay occurs, so that their momenta cannot cancel out to equal the initial momentum of zero.

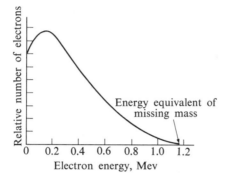

FIG. 10–1. The distribution of electron energies found in the beta decay of $_{83}Bi^{210}$. The maximum electron energy is equal to the energy equivalent of the mass lost by the decaying nucleus minus the electron mass.

A third difficulty concerns angular momentum. The spin quantum number of the neutron, the proton, the electron, and the positron is in every case one-half. The conversion of a neutron into a proton and an electron or of a proton into a neutron and a positron therefore leaves an angular momentum discrepancy of $\frac{1}{2}\,h/2\pi$.

The process of beta decay is the first one we have encountered in which the conservation laws of energy, momentum, and angular momentum do not seem to hold. To account for the above discrepancies without abandoning three of the most fundamental and otherwise well-established physical principles, the existence of a new particle was postulated, the *neutrino*, symbol ν (Greek letter nu). The neutrino has no electric charge and no mass, but is able to possess both energy and momentum and has an intrinsic spin of one-half. (Lest this seem unlikely, we might reflect that the photon, also massless, has energy, momentum, and angular momentum. The neutrino is *not* a photon, however, but is an entirely different entity.) According to the neutrino theory, an electron and neutrino are simultaneously emitted in beta decay, as in Fig. 10–2, which permits energy and momentum to be conserved.

For a quarter of a century the existence of the neutrino was accepted despite the absence of any direct empirical evidence in support of it.

Finally, in 1956, an experiment was performed in which a nuclear reaction that, in
theory, could only be caused by a neutrino,
was actually found to take place. The difficulty to be overcome in detecting neutrinos is their exceedingly feeble interaction
with other particles. As we learned in the
previous chapter, solar energy originates in
nuclear reactions that take place in its interior. Beta decays occur at several stages
in the sequences of reactions that participate in the conversion of hydrogen to helium, and a vast number of neutrinos is
accordingly produced within the sun. Because neutrinos traverse matter freely, al-

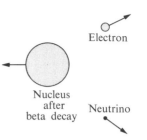

FIG. 10-2. An electron and a
neutrino are simultaneously emitted
in the beta decay of a nucleus, which
makes possible the conservation of
both energy and momentum in the
process.

most all of these neutrinos escape into space and take with them six to
eight percent of the total power generated by the sun. The flux of neutrinos from the sun is such that every cubic inch on the earth contains
perhaps 100 neutrinos at any instant! In its operation a nuclear reactor
also creates a great many neutrinos, and more than 10^{16} neutrinos per
second may emerge from each square meter of the shielding around a
reactor. A neutrino striking a proton has a small probability of inducing
the reaction

$$\nu + \mathrm{p} \rightarrow \mathrm{n} + \mathrm{e}^+ \tag{10-1}$$

in which a neutron and positron are created. By placing a sensitive
detecting chamber containing hydrogen near a nuclear reactor, the simultaneous appearance of a neutron and a positron could be registered each
time the above reaction occurred. Calculations were made initially
of how many such reactions per second should occur based on the known
properties of the detecting apparatus and on the theoretical properties of
the neutrino. When this reaction rate was actually found, there was no
doubt that neutrinos indeed exist.

   As we shall learn in a later section of this chapter, another significant
process in which neutrinos are emitted is the decay of charged π-mesons.
These particles have masses intermediate between those of electrons and
of protons, and may have charges of $+e$ or $-e$. In free space π-mesons
decay into μ-mesons and neutrinos; a neutrino is associated with the
decay of a positive π-meson and an antineutrino with the decay of a
negative π-meson. Are these neutrinos the same particles as those
emitted in ordinary beta decay?

   The identity of the neutrinos involved in meson decay was finally
established in 1962. In this experiment π-mesons were produced by bombarding a metal target with high-energy protons. The meson decays

liberated neutrinos, and the interactions of these neutrinos with matter was studied. The only inverse reactions found led to the production of μ-mesons; no electrons whatever were created. Hence the neutrinos set free in meson decay are different from those set free in beta decay. The symbols for the two kinds of neutrino are respectively ν_e and ν_μ.

10–2 ANTIPARTICLES

Another recent experimental discovery of a particle whose existence had been predicted theoretically decades ago is that of the negative proton, or *antiproton*, whose symbol is $\bar{p}$. This is a particle with the same properties as the proton except that it has a negative electric charge. The existence of antiprotons was predicted largely on the basis of symmetry arguments: since the electron has a positive counterpart in the positron, why should the proton not have a negative counterpart as well? Actually, as sophisticated theories show, this is an excellent argument, and few physicists were surprised when the antiproton was actually found.

The reason positrons and antiprotons are so difficult to find is that they are readily *annihilated* upon contact with ordinary matter. When a positron is in the vicinity of an electron, they attract one another electrostatically, come together, and then both vanish simultaneously, with their missing mass appearing in the form of two gamma-ray photons (Fig. 10–3):

$$e^+ + e^- \rightarrow \gamma + \gamma. \tag{10–2}$$

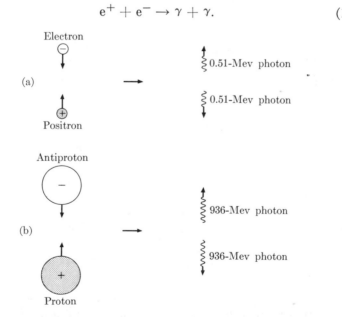

FIG. 10–3. The mutual annihilation of (a) an electron and a positron, and (b) a proton and an antiproton.

The total mass of the two particles is the equivalent of 1.02 Mev, and so each photon has an energy of 0.51 Mev. (Their energies must be equal and they must be emitted in opposite directions in order that momentum be conserved.) While the similar reaction

$$p + \bar{p} \rightarrow \gamma + \gamma \qquad (10\text{–}3)$$

can occur when a proton and antiproton undergo annihilation, it is more usual for the vanished mass to reappear in the form of several π-mesons, particles which we shall consider in Section 10–3.

The reverse of annihilation can also take place, with the electromagnetic energy of a photon materializing into a positron and an electron or, if it is energetic enough, into a proton and an antiproton (Fig. 10–4). This phenomenon is known as *pair production,* and requires the presence of a nucleus in order to happen. Any photon energy in excess of the amount required to provide the mass of the created particles (1.02 Mev for a positron-electron pair, 1872 Mev for a proton-antiproton pair) appears as kinetic energy.

Antineutrons (symbol $\bar{n}$) and antineutrinos (symbol $\bar{\nu}_e$ and $\bar{\nu}_\mu$) have also been identified. Antineutrons can be detected through their mutual annihilation with neutrons, while more indirect, though equally definite, evidence supports the existence of antineutrinos. The antineutrino differs from the neutrino in that, while the spin axes of both are parallel to their

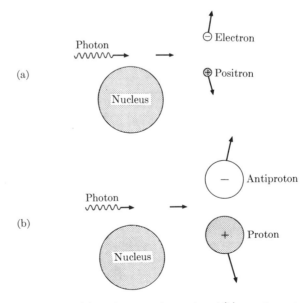

FIG. 10–4. The production of (a) an electron-positron pair and (b) a proton-antiproton pair by the materialization of sufficiently energetic photons. Pair production can occur only in the presence of nucleus.

directions of motion, the spin of the former is clockwise and that of the latter is counterclockwise when viewed from behind. A moving neutrino may be thought of as resembling a left-handed screw, and a moving anti-neutrino as resembling a right-handed screw. An antineutrino is released during a beta decay in which an electron is emitted, and a neutrino is released during a beta decay in which a positron is emitted. Thus the fundamental equations of beta decay are

$$p \rightarrow n + e^+ + \nu_e, \tag{10-4}$$

$$n \rightarrow p + e^- + \bar{\nu}_e. \tag{10-5}$$

Annihilation and pair production are consequences of the facts that matter is a form of energy and that conversions from matter to energy and from energy to matter are no more improbable than conversions from, say, gravitational potential energy to kinetic energy. Relativity and quantum effects underlie the reality of our experience as much as more familiar ones such as the kinetic nature of heat, and within the past fifteen or twenty years the full realization of their significance has profoundly affected the lives of all of us.

10–3 MESONS

If nuclear forces indeed arise from the exchange of mesons between nearby nuclei, as we mentioned in Chapter 8, it should be possible to detect their existence by imparting enough energy to a nucleus for·con-version into meson mass without affecting the nuclear mass itself. We can estimate the meson mass from the uncertainty principle,

$$\Delta E \, \Delta t \geq h. \tag{10-6}$$

We assume that the temporary energy discrepancy ΔE is of the same magnitude as the rest energy mc^2 of the meson, and that the meson speed is of the same magnitude as the speed of light c as it travels from one nucleon to another. The time Δt the meson spends between creation and absorption cannot be greater than R/c, where R is the maximum dis-tance that can separate interacting nucleons. We therefore have

$$\Delta E \, \Delta t \approx h,$$

$$mc^2 \times \frac{R}{c} \approx h,$$

$$m \approx \frac{h}{Rc}. \tag{10-7}$$

Nuclear forces have a range of about 2×10^{-15} meter. When we sub-titute

$$R = 2 \times 10^{-15} \text{ meter}$$

into Eq. (10–7) we find a predicted meson mass of 10^{-27} kg, about a thousand times the mass of the electron. Of course, the above calculation is hardly a precise one, but, if Yukawa's theory has any validity, the mesons he postulated should have masses between those of electrons and those of nucleons.

Not long after Yukawa's work, charged particles of about the right mass were experimentally discovered in the cosmic radiation. Their discovery was not unexpected because a sufficiently energetic nuclear collision should be able to liberate mesons by providing enough energy to create them without violating conservation of energy, and nuclear collisions between fast cosmic-ray protons from space and oxygen and nitrogen nuclei occur constantly in the atmosphere. However, these particles did not behave at all in the way they were expected to behave. Far from strongly interacting with nuclei, as Yukawa's mesons were supposed to do, they barely interacted at all. Instead of being absorbed in at most a meter of earth, they penetrated thousands of meters into the ground. Finally, in 1947, the explanation for their peculiar behavior was found. The weakly interacting mesons, known as μ-mesons (μ is the Greek letter mu; μ-mesons are often called *muons*) and found so profusely in cosmic rays at sea level, are not the direct products of nuclear collisions, but are secondary particles that result from the decay of the Yukawa mesons. The meson postulated by Yukawa is known as the π-meson (or *pion*), and when outside a nucleus it usually decays in 2.54×10^{-8} sec into a μ-meson and a neutrino plus kinetic energy. That is,

$$\pi^+ \rightarrow \mu^+ + \nu_\mu, \tag{10–8}$$

$$\pi^- \rightarrow \mu^- + \bar{\nu}_\mu. \tag{10–9}$$

(A small percentage of charged π-mesons decay directly into an electron and a neutrino.) The π-meson is observed to have the properties Yukawa attributed to it, and may be loosely thought of as the "glue" that holds nuclei together so strongly. Charged π-mesons have a mass of 273 m_e. Neutral π-mesons also exist; their mass is 264 m_e, a little less than that of their charged counterparts, and they decay in about 10^{-16} sec into two gamma-ray photons:

$$\pi^0 \rightarrow \gamma + \gamma. \tag{10–10}$$

The brief lifetimes of π-mesons are responsible for the delay in their discovery. In the past decade, high-energy accelerators have been built whose particles are able to violently disrupt nuclei with the creation of many π-mesons; considerable progress has been made in our understanding of these mesons and their behavior. The π^--meson is the antiparticle of the π^+-meson, and the neutral π^0-meson is its own antiparticle.

The μ-meson, whose mass is 207 m_e, is less well understood than the π-meson. The relatively long μ-meson lifetime of 2.22×10^{-6} sec and their feeble interaction with matter accounts for the penetrating ability of these mesons. In contrast to the π-meson, which fits in well with the theory of nuclear forces, the μ-meson seems to have no particular function in the scheme of things; why such a particle should exist is one of the current problems of theoretical physics. The μ^+- and μ^--mesons decay into positrons and electrons together with a neutrino and an antineutrino in each case:

$$\mu^+ \to e^+ + \nu_e + \bar{\nu}_e, \tag{10–11}$$

$$\mu^- \to e^- + \nu_e + \bar{\nu}_e. \tag{10–12}$$

The μ^+-meson is the antiparticle of the μ^--meson. There is no neutral μ-meson.

10–4 STRANGE PARTICLES

Not long after the discovery of the π-meson a succession of other, quite unpredicted elementary particles came to light. Even though a good deal is now known about these new particles, they may still be called by their original name of *strange particles* with some justice. It is convenient to divide strange particles into two classes, *K-mesons*, whose masses are all almost exactly $967\ m_e$, and *hyperons*, whose masses exceed the mass of the proton. Both classes of strange particles, like π-mesons, are found as products of energetic nuclear collisions, and are unstable.

Charged K-mesons may decay in a variety of ways after a mean lifetime of 1.22×10^{-8} sec. The K^+-meson has been observed to have the following six decay schemes, listed in the order of probability:

$$
\begin{aligned}
K^+ &\to \mu^+ + \nu_\mu \\
&\to \pi^+ + \pi^0 \\
&\to \pi^+ + \pi^+ + \pi^- \\
&\to \pi^+ + \pi^0 + \pi^0 \\
&\to e^+ + \pi^0 + \nu_e \\
&\to \mu^+ + \pi^0 + \nu_\mu.
\end{aligned}
\tag{10–13}
$$

The K^--meson, the antiparticle of the K^+-meson, has similar decay possibilities. Two different versions of the neutral K-meson are known. The K_1^0-meson decays in 10^{-10} sec in either of these ways:

$$
\begin{aligned}
K_1^0 &\to \pi^+ + \pi^- \\
&\to \pi^0 + \pi^0.
\end{aligned}
\tag{10–14}
$$

The K_2^0-meson has the much longer mean lifetime of 6×10^{-8} sec and six possible decay schemes:

$$K_2^0 \rightarrow \pi^+ + \pi^- + \pi^0$$
$$\rightarrow \pi^0 + \pi^0 + \pi^0$$
$$\rightarrow \pi^- + \mu^+ + \nu_\mu$$
$$\rightarrow \pi^+ + \mu^- + \bar{\nu}_\mu$$ (10–15)
$$\rightarrow \pi^- + e^+ + \nu_e$$
$$\rightarrow \pi^+ + e^- + \bar{\nu}_e.$$

Unlike the π^0-meson, which is its own antiparticle, an anti-K^0-meson, the $\bar{K}^0$-meson, is believed to exist.

There are three known families of hyperons, all of whose members are heavier than nucleons and all of which have a nucleon as one of their decay products. There is only a single kind of Λ-hyperon (Λ is the Greek capital letter lambda), a neutral particle whose mass is $2182\ m_e$ that decays in an average of 2.5×10^{-10} sec into either a proton and π^--meson or into a neutron and a π^0-meson, the latter alternative being the less likely:

$$\Lambda^0 \rightarrow p^+ + \pi^-$$
$$\rightarrow n^0 + \pi^0.$$ (10–16)

The $\bar{\Lambda}^0$-hyperon is the antiparticle of the Λ^0-hyperon, and decays into antiparticle counterparts of the above products.

The Σ-hyperons may be neutral or may have either charge (Σ is the Greek capital letter sigma). The Σ^+-hyperon has a mass of $2327\ m_e$, a mean lifetime of 0.8×10^{-10} sec, and an equal probability of decaying into a proton and a π^0-meson or into a neutron and a π^+-meson:

$$\Sigma^+ \rightarrow p^+ + \pi^0$$
$$\rightarrow n^0 + \pi^+.$$ (10–17)

Only one decay mode is known for the Σ^--hyperon, whose mass is $2341\ m_e$ and whose mean lifetime is 1.6×10^{-10} sec:

$$\Sigma^- \rightarrow n^0 + \pi^-.$$ (10–18)

The Σ^0-hyperon has a mass of $2329\ m_e$ and a mean lifetime of less than 10^{-12} sec. The Σ^0-hyperon decays into a Λ^0-hyperon and a gamma ray:

$$\Sigma^0 \rightarrow \Lambda^0 + \gamma.$$ (10–19)

The $\bar{\Sigma}^+$-, $\bar{\Sigma}^-$-, and $\bar{\Sigma}^0$-hyperons are antiparticles of the Σ^+-, Σ^--, and Σ^0-hyperons.

The third family of hyperons has two members, the Ξ^--hyperon and the Ξ^0-hyperon (Ξ is the Greek capital letter xi). Their masses are about 2583 m_e, and their lifetimes are in the range 10^{-9} to 10^{-10} sec. Both decay into a Λ^0-hyperon and a π-meson,

$$\Xi^- \rightarrow \Lambda^0 + \pi^-, \tag{10–20}$$

$$\Xi^0 \rightarrow \Lambda^0 + \pi^0, \tag{10–21}$$

and both have appropriate antiparticles.

The fourth type of hyperon, the Ω^--hyperon, is the heaviest, with a mass of 3290 m_e. Its possible decay schemes are

$$\Omega^- \rightarrow \Lambda^0 + K^- \tag{10–22}$$

$$\rightarrow \Xi^0 + \pi^- \tag{10–23}$$

$$\rightarrow \Xi^- + \pi^0. \tag{10–24}$$

28–5 CATEGORIES OF ELEMENTARY PARTICLES

In all, about 100 elementary particles have been discovered, of which we have been introduced to no less than 25 in this chapter. Such abundance where scarcity had been expected inspired considerable interest among physicists in the past decade, with the result that partial success has been attained in understanding these entities. But there is still a long way to go before this fundamental puzzle is solved.

Elementary particles fall naturally into four categories, namely, *photons, leptons, mesons,* and *baryons.* The members of each category share certain basic properties despite differences among themselves in other properties. Photons, as we know, are quanta of electromagnetic energy; they are stable and have zero rest mass. The category of leptons includes neutrinos, electrons, and μ-mesons. Leptons exhibit very weak interactions with nuclei and have a spin of one-half. (We note that in this classification, μ-mesons are considered leptons rather than mesons.) The meson category includes both π- and K-mesons. These particles interact quite strongly with nuclei and all have a spin of zero. Baryons include nucleons and hyperons; they also interact strongly with nuclei, but have spins of one-half. Table 10–1 lists 34 of the longest-lived elementary particles together with some of their characteristic properties; the others exist only for the briefest of instants before decaying.

While it is obvious from Table 10–1 that the various elementary particles fit into the indicated categories on the basis of their masses and spins, experimental evidence of an unusual kind further substantiates this division. Let us assign all leptons the special quantum number $L = +1$, all antileptons $L = -1$, and all other particles $L = 0$. It is found that in every known process involving leptons, including their production, decay, or interaction with other particles, the total value of L remains constant. Also, let us assign all baryons the special quantum number $B = +1$, all antibaryons $B = -1$, and all other particles $B = 0$.

TABLE 10–1

Some elementary particles*

Name	Particle	Antiparticle	Mass m_e	Spin	L	B	S	Mean lifetime (sec)	Category
Photon	γ	(γ)	0	1	0	0	0	Stable	Photon
Neutrino	ν_e, ν_μ	$\bar{\nu}_e, \bar{\nu}_\mu$	0	$\frac{1}{2}$	+1	0	0	Stable	Leptons
Electron	e^-	e^+	1	$\frac{1}{2}$	+1	0	0	Stable	
Mu-meson	μ^-	μ^+	207	$\frac{1}{2}$	+1	0	-1	2.22×10^{-6}	
Pi-meson	π^+	π^-	273	0	0	0	0	2.54×10^{-8}	Mesons
	π^0	(π^0)	264	0	0	0	0	10^{-16}	
K-meson	K^+	K^-	967	0	0	0	+1	1.22×10^{-8}	
	K^0	$\overline{K}^0$	967	0	0	0	+1	$10^{-10}; 6 \times 10^{-8}$	
Proton	p	$\bar{p}$	1836	$\frac{1}{2}$	0	+1	0	Stable	Baryons
Neutron	n^0	$\bar{n}^0$	1839	$\frac{1}{2}$	0	+1	0	1.11×10^3	
Lambda-hyperon	Λ^0	$\overline{\Lambda}^0$	2182	$\frac{1}{2}$	0	+1	-1	2.5×10^{-10}	
Sigma-hyperon	Σ^+	$\overline{\Sigma}^-$	2327	$\frac{1}{2}$	0	+1	-1	0.8×10^{-10}	
	Σ^-	$\overline{\Sigma}^+$	2341	$\frac{1}{2}$	0	+1	-1	1.6×10^{-10}	
	Σ^0	$\overline{\Sigma}^0$	2329	$\frac{1}{2}$	0	+1	-1	10^{-12}	
Xi-hyperon	Ξ^-	$\overline{\Xi}^-$	2583	$\frac{1}{2}$	0	+1	-2	$10^{-9} - 10^{-10}$	
	Ξ^0	$\overline{\Xi}^0$	2583	$\frac{1}{2}$	0	+1	-2	$10^{-9} - 10^{-10}$	
Omega-hyperon	Ω^-	$\overline{\Omega}^+$	3290	$\frac{3}{2}$	0	+1	-3	10^{-10}	

* Decay schemes given in text.

It is found that in every known process involving baryons, the total value of B remains constant. A simple example is the beta decay of the neutron:

$$n^0 \rightarrow \quad p^+ \ + \ e^- \ + \ \bar{\nu}_e \qquad (10\text{--}25)$$

$$L = \quad 0 \qquad 0 \qquad +1 \qquad -1$$

$$B = +1 \qquad +1 \qquad 0 \qquad 0$$

The usual conservation laws of energy, momentum, angular momentum, and charge, plus the new (and still rather mysterious) conservation laws of lepton and baryon numbers, help in understanding why certain processes involving elementary particles occur and why others do not occur. For instance, the free neutron can decay according to Eq. (10–25) and still obey the above conservation laws, but in no other way. The free proton, which has less mass than the neutron, cannot decay at all: energy conservation prohibits it from decaying with any other baryon as a product, and baryon number conservation prohibits it from decaying into any lighter particles. The other stable particles are also prohibited from decaying by one or more of the above laws when they are in free space.

Introducing the quantum numbers L and B did not clarify all aspects of elementary particle behavior. Three of the still outstanding problems were the following:

(1) Some heavy particles decay into lighter ones plus gamma rays while others do not. For example, the Σ^0-baryon decays into a Λ^0-baryon and a gamma ray,

$$\Sigma^0 \rightarrow \Lambda^0 + \gamma,$$

while the Σ^+-baryon never decays in the analogous manner

$$\Sigma^+ \not\rightarrow p^+ + \gamma.$$

(2) Physical processes that release large amounts of energy almost always occur faster than processes involving lesser amounts. However, relatively long-lived particles are known whose decays evolve tens or hundreds of Mev's.

(3) Strange particles never come into being singly, but always two or more at a time.

Observations such as these led to the introduction of a further quantum number, the *strangeness number S*. Table 10–1 shows the values of S that are assigned to the various elementary particles. Only the photon and π^0-meson have $L = B = S = 0$. Since these particles also have no charge, it is impossible to distinguish between them and their antiparticles, and the photon and π^0-meson are accordingly regarded as their own antiparticles. Before we discuss how the strangeness number may be interpreted, it will be necessary to look into another aspect of elementary particle physics.

10–6 FUNDAMENTAL INTERACTIONS

Elementary particles can interact with one another in four different ways, and these interactions are responsible for all of the physical processes in the universe. Feeblest is the gravitational interaction, which follows the inverse-square law discovered by Newton. Next is the *weak interaction*, characteristic of leptons, that acts between leptons and either mesons, baryons, or other leptons (besides any electromagnetic forces that may also be present). The weak interaction is responsible for reactions involving neutrinos, such as beta decays. Of greater strength than gravitational and weak interactions are the electromagnetic interactions between all charged particles and also between those with electric or magnetic moments. Most powerful of the four is the *strong interaction* between mesons, between baryons, and between mesons and baryons that gives rise to nuclear forces.

The relative magnitudes of the strong, electromagnetic, weak, and gravitational interactions are in the ratios $1:10^{-2}:10^{-14}:10^{-39}$. To be sure, the ranges of the forces produced by these interactions are quite different: the nuclear force between two nucleons vastly exceeds the gravitational force between them when they are 10^{-15} m apart, but when they are one meter apart the situation is reversed. Nuclear structure is a consequence of the properties of the strong interaction, while atomic structure is a consequence of those of the electromagnetic interaction. Because bulk matter is electrically neutral and the strong and weak forces are severely limited in their range, gravitational forces, negligible on a small scale, become all-important on a large scale. The role of the weak force in the structure of matter seems to be restricted to causing beta decays that adjust the neutron:proton ratios of nuclei to stable values.

In all processes involving strong and electromagnetic interactions, the strangeness number S is conserved. Thus the decay

$$\Sigma^0 \;\rightarrow\; \Lambda^0 + \gamma$$
$$S = -1 \qquad -1 \qquad 0$$

that we mentioned earlier conserves S and actually occurs, and the apparently similar decay

$$\Sigma^+ \;\nrightarrow\; p^+ + \gamma$$
$$S = -1 \qquad 0 \qquad 0$$

does not conserve S and has never been found to occur. The multiple production of strange particles in high-energy nuclear collisions is necessary in order to conserve S, since such collisions involve strong interactions. The relatively long lifetimes of all unstable particles except the π^0-meson and Λ^0-hyperon follow from the assumption that weak inter-actions are also characteristic of mesons and baryons as well as leptons, though normally dominated by strong or electromagnetic interactions in the former cases. With strong or electromagnetic decay processes im-possible except for the π^0-meson and Λ^0-hyperon because S must be con-served in such processes, only the weak interaction is available for particle decay. Events governed by weak interactions are very slow, in agree-ment with observation. However, even in weak interactions S does not change by more than $+1$ or -1. Thus the Ξ^--hyperon does not decay directly into a neutron in the single step

$$\Xi^- \quad \nrightarrow \quad n^0 + \pi^-$$
$$S = -2 \qquad 0 \quad \ 0$$

but instead requires two separate decays:

$$\Xi^- \quad \rightarrow \quad \Lambda^0 + \pi^- \qquad \qquad \Lambda^0 \rightarrow n^0 + \pi^0$$
$$\text{and}$$
$$S = -2 \qquad \ -1 \quad \ 0 \qquad \qquad S = -1 \quad \ \ 0 \quad \ 0$$

Despite the elegance with which the introduction of the quantum numbers L, B, and S permits us to organize our knowledge of elementary particle phenomena, much remains to be understood about the origins of these numbers and their relation to the four fundamental interactions. The study of elementary particles is accordingly one of the most active branches of current scientific research.

PROBLEMS

1. State which of the following are spontaneously emitted by radioactive nuclei, and which are not: electrons, positrons, neutrinos, protons, antiprotons, neu-trons, mesons, alpha particles, electromagnetic waves.

2. How much energy must a gamma-ray photon have if it is to materialize into a proton-antiproton pair with each particle having a kinetic energy of 10 mev?

3. How much energy (in ev) must a gamma-ray photon have if it is to mate-rialize into a neutron-antineutron pair? Is this more or less than that re-quired to form a proton-antiproton pair?

4. A 1-Mev positron collides head on with a 1-Mev electron, and the two are annihilated. What is the energy and wavelength of each of the resulting gamma-ray photons?

5. One proton strikes another, and the reaction

$$p + p \rightarrow n + p + \pi^+$$

takes place. What is the minimum energy the incident proton must have had?

6. Find the energy of each of the gamma-ray photons produced in the decay of the neutral pi-meson. Why must their energies be the same?

7. Why does the neutron not decay into an electron, a positron, and a neutrino?

8. Why does the Λ-hyperon not decay into a π^+- and a π^--meson?

9. A proton can interact with a μ^--meson to form a neutron and another particle. What must this particle be?

10. According to the theory of the continuous creation of matter, the evolution of the universe can be traced to the spontaneous appearance of neutrons in free space. Which conservation laws would this process violate?

APPENDIXES

Appendix A

KINETIC–MOLECULAR THEORY OF MATTER

A–1 MOLECULAR THEORY OF MATTER

The idea that matter is not infinitely divisible, that all substances are composed of characteristic individual particles, is an ancient one. The ultimate particles of any substance are called *molecules*. Although molecules may be further broken down, when this happens they no longer are representative of the original substance. The molecules of a compound consist of the *atoms* of its constituent elements joined together in a definite ratio. Thus each molecule of water contains two hydrogen atoms and one oxygen atom. While the ultimate particles of elements are atoms, many elemental gases consist of molecules rather than atoms. Oxygen molecules, for instance, contain two oxygen atoms each.. The molecules of other gases, such as helium and argon, are single atoms. Figure A–1 shows schematically the composition of some common molecules.

The masses of atoms and molecules are usually expressed in units such that the most abundant type of carbon atom has a mass of precisely 12. (The existence of different kinds of atoms, known as *isotopes*, of the same chemical species is discussed in Chapter 8.) It is customary to refer to masses in these units as *atomic weights* and *molecular weights*. Appendix D contains a list of the atomic weights of the elements; we note that ordinary carbon, some of whose atoms are heavier than the most abundant kind, has an atomic weight of 12.01. If we know the composition of a compound, we can determine the corresponding molecular weight. The carbon atom has an actual mass of 1.992×10^{-26} kg, and so

$$1 \text{ atomic weight unit} = \frac{1.992 \times 10^{-26}\,\text{kg}}{12} = 1.660 \times 10^{-27}\,\text{kg}.$$

The molecular weight of a compound is the total of the atomic weights of the elements it contains multiplied respectively by the numbers of their atoms in each molecule of the compound. Thus the molecular

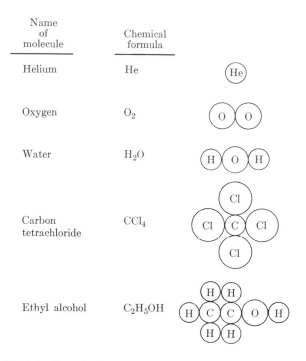

Name of molecule	Chemical formula
Helium	He
Oxygen	O_2
Water	H_2O
Carbon tetrachloride	CCl_4
Ethyl alcohol	C_2H_5OH

FIG. A–1. The molecular compositions of several common substances.

weight of gaseous oxygen is **32**, since its formula is O_2 and the atomic weight of oxygen is **16**. To find the molecular weight of the more complicated compound, ethyl alcohol (C_2H_5OH), we proceed as follows:

$$2\ C = 2 \times 12.01\ = 24.02$$
$$6\ H = 6 \times 1.008 =\ 6.05$$
$$1\ O = 1 \times 16.00\ = 16.00$$
$$\text{total} =\ 46.07$$

The molecular weight of ethyl alcohol is **46.07**.

A *kilomole* (abbreviated *kmole*) of any element or compound is that quantity of it whose mass is equal to its atomic or molecular weight expressed in kilograms. One kmole of carbon therefore has a mass of 12.01 kg, and one kmole of ethyl alcohol has a mass of 46.07 kg. The usefulness of the kmole arises from the fact that a kmole of any compound contains exactly as many molecules as a kmole of any other compound. The kmole is widely used in chemistry as a measure of quantity, instead of actual mass, because the primary interest of the chemist is usually in the relative numbers of atoms and molecules that react together. A kmole of an element is sometimes referred to as a *kilogram-atom* of that element; the *gram-mole* (or simply *mole*) and *gram-atom* are smaller than the kmole and kilogram-atom by a factor of 10^3, since there are 10^3 grams in a kilogram.

A–2 AVOGADRO'S NUMBER

The number of molecules in a kmole is a universal constant known as *Avogadro's number*, symbol N_0. Its value is

$$N_0 = 6.025 \times 10^{26} \text{ molecules/kmole.}$$

Avogadro's number can be determined in a variety of ways. A particularly simple one is based upon the spreading of an oil film on calm water, where we assume that at its maximum extension the film is exactly one molecule thick. A convenient substance for this purpose is oleic acid, a constituent of most animal fats and vegetable oils. One milligram (10^{-6} kg) of oleic acid placed on a water surface covers an area of about 1.4 m^2 under the most favorable conditions. The density of oleic acid is 895 kg/m^3, so that 1 mgm occupies a volume of $1.12 \times 10^{-9} \text{ m}^3$. Hence the thickness of the film is

$$a = \frac{1.12 \times 10^{-9} \text{ m}^3}{1.4 \text{ m}^2} = 8.0 \times 10^{-10} \text{ m.}$$

Under the assumption that this length is characteristic of all the dimensions of the oleic acid molecule, the volume occupied by each molecule is

$$a^3 = 5.1 \times 10^{-28} \text{ m}^2.$$

From the formula of oleic acid, $CH_3(CH_2)_7CH{=}CH(CH_2)_7COOH$ (the reason for writing the formula in this manner is given in Chapter **24**), we find that its molecular weight is 282.5. The volume per kmole is therefore

$$\frac{\text{volume}}{\text{kmole}} = \frac{\text{molecular weight}}{\text{density}} = \frac{282.5 \text{ kg/kmole}}{895 \text{ kg/m}^3} = 0.316 \text{ m}^3/\text{kmole.}$$

Since we suppose that each molecule has a volume of $5.1 \times 10^{-28} \text{ m}^3$, the number of molecules per kmole of oleic acid is

$$N_0 = \frac{0.316 \text{ m}^3/\text{kmole}}{5.1 \times 10^{-28} \text{ m}^3/\text{molecule}} = 6.2 \times 10^{26} \text{ molecules/kmole.}$$

Thus a quite primitive experiment yields a value for Avogadro's number not far from its actual value of 6.025×10^{26} molecules/kmole. More accurate methods of determining N_0 involve x-ray diffraction and electrolysis, as mentioned in Chapters **6** and **7**.

The specimens of matter we normally deal with contain immense numbers of atoms and molecules, and actually counting them is out of the question. However, if we know the mass of a sample of a particular element or compound, we can find the number N of atoms or molecules it contains by using the definition of the kmole together with Avogadro's number. Since

$$N = nN_0, \tag{A-1}$$

where n is the number of kmoles in the sample and N_0 is Avogadro's number, we have

$$N = \frac{\text{mass of sample}}{\text{molecular weight}} \times N_0. \tag{A-2}$$

A kilogram of ethyl alcohol, for instance, contains

$$N = \frac{1 \text{ kg}}{46.07 \text{ kg/kmole}} \times 6.025 \times 10^{26} \text{ molecules/kmole}$$

$$= 1.34 \times 10^{25} \text{ molecules.}$$

It is an experimental fact that under identical conditions of pressure and temperature, equal volumes of gases contain equal numbers of molecules. The reason for this is that gas molecules themselves have negligible volumes compared with the sizes of their containers; gas molecules fill their containers and exert pressure on their walls because of their random motions, an idea we shall explore in detail in the remainder of this chapter. Under the standard conditions of atmospheric pressure and 0°C, often referred to by the abbreviation STP, one kmole of any gas occupies a volume of 22.4 m³. Let us use this observation to evaluate the constant factor in the ideal gas law,

$$\frac{pV}{T} = \text{constant.}$$

At STP, $p_0 = 1.013 \times 10^5 \text{ n/m}^2$, $T_0 = 273°\text{K}$, and $V_0 = 22.4 \text{ m}^3/\text{kmole}$ of gas present. If there are n kmoles in the sample, its volume is nV_0 at STP. Hence, in general,

$$\frac{pV}{T} = \frac{np_0V_0}{T_0} = nR, \tag{A-3}$$

where the quantity

$$R = \frac{p_0V_0}{T_0} = \frac{1.013 \times 10^5 \text{ n/m}^2 \times 22.4 \text{ m}^3/\text{kmole}}{273°\text{K}}$$

$$= 8.32 \times 10^3 \text{ j/kmole} \cdot °\text{K}$$

$$= 1.99 \text{ kcal/kmole} \cdot °\text{K}$$

is known as the *universal gas constant*. The complete expression of the ideal gas law is therefore

$$pV = nRT. \qquad (A\text{--}4)$$

Though at first glance it is just another fact to add to our collection, the presence of 6.025×10^{26} molecules in $22.4 \, m^3$ of any gas at STP is surely one of the most remarkable conclusions we have yet encountered. If a cubic centimeter of air at STP—a thimbleful—were to be divided equally among all the three billion people on the earth, each would receive 200 billion molecules! There are about as many molecules in an average breath of air as there are breaths in the entire atmosphere, so that, as James Jeans has said, "if we assume that the last breath of, say, Julius Caesar, has by now become thoroughly scattered through the atmosphere, then the chances are that each of us inhales one molecule of it with every breath we take."

A considerable amount of experimentation and ingenious reasoning had to be carried out before the reality of atoms and molecules became definitely established. Although we will not go into the full story of the molecular theory of matter, a large part of which involves chemistry, we shall show that it can account for the ideal gas law. We shall also discuss briefly how the physical properties of solids and liquids, and the deviations of actual gases from the ideal gas, fit into the molecular theory.

A–3 KINETIC THEORY OF GASES

According to the assumptions of the *kinetic theory of gases,* a development of the molecular theory of matter, a gas consists of a great many tiny individual molecules that do not interact with one another except when collisions occur. The molecules are supposed to be far apart compared with their dimensions and to be in constant motion, incessantly hurtling to and fro as in Fig. A–2, being kept from escaping into space only by the solid walls of a container (or, in the case of the earth's atmosphere, by gravity). A natural consequence of the random motion and large molecular separation is the tendency of a gas to completely fill its container and to be readily compressed or expanded. In a solid, on the other hand, the molecules are close together, and mutual attractive and repulsive forces hold them in place to provide the solid with its characteristic rigidity. In a liquid the intermolecular forces are sufficient to keep the volume of the liquid constant; however, they are not

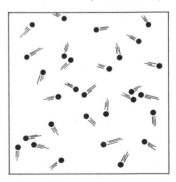

FIG. A–2. The pressure a gas exerts originates in the impacts of its molecules, which are in constant, random motion.

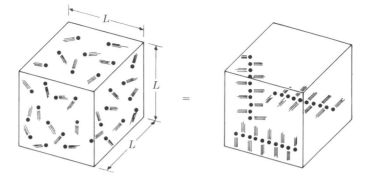

FIG. A-3. A simplified model of a gas.

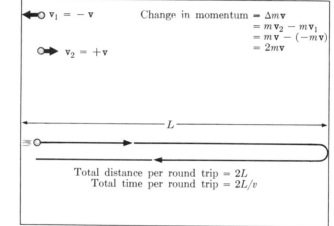

FIG. A-4

strong enough to prevent adjacent molecules from sliding past one another, which results in the ability of liquids to flow.

Let us see how the kinetic theory of gases accounts for the ideal gas law. At first glance it may not seem that there *is* any straightforward connection between them. But it is also clear that if the kinetic theory is to have any meaning at all, it must yield the same behavior that we obtain from experiment. To bridge the gap between the microscopic picture of a gas as an aggregate of molecules in random motion and the macroscopic picture of a gas as a continuous fluid with certain physical properties, we shall use the laws of mechanics. The derivation is worth following in detail, since it illustrates how a theory whose assumptions are not based upon direct observation may be compared with data obtained by such observation.

We begin with a model situation: a box L long on each side filled with N identical molecules each having the same mass m. While these molecules are actually traveling about in all directions, the effects of their collisions with the walls of the box are the same as if one-third of them were moving back and forth between each pair of opposite walls (Fig. A–3). We therefore imagine, for the sake of convenience, that one-third of our N molecules bounce between the top and bottom walls of the box, one-third between the front and rear walls, and one-third between the right and left walls.

Next we consider what happens when a molecule strikes one of the walls. As in Fig. A–4, it approaches the wall with the velocity $-\mathbf{v}$, so that its initial momentum is $-m\mathbf{v}$, and bounces off with the velocity $+\mathbf{v}$ (same speed but opposite direction), so that its final momentum is $+m\mathbf{v}$. Hence the molecule has experienced a *change in momentum* of

$$\Delta m\mathbf{v} = m\mathbf{v}_2 - m\mathbf{v}_1 = m\mathbf{v} - (-m\mathbf{v}) = 2m\mathbf{v}.$$

The distance between the walls of the box is L, and it takes a molecule whose speed is v the time

$$\Delta t = t_2 - t_1 = \frac{2L}{v}$$

to make the round trip from one wall to the opposite one and back again. (The time required is the total distance divided by the speed, and here the total distance is $2L$.) The change in the molecule's momentum when it strikes the wall is equal in magnitude to the impulse

$$\mathbf{F}(t_2 - t_1) = \mathbf{F}\,\Delta t$$

it exerts on the wall, where $\mathbf{F}$ is the average force during the period of time Δt. Hence

$$\mathbf{F}\,\Delta t = \Delta m\mathbf{v},$$

and the magnitude of the average force on the wall caused by the successive impacts of the molecule we are considering is

$$F = \frac{\Delta mv}{\Delta t} = \frac{2mv}{2L/v} = \frac{mv^2}{L}. \tag{A–5}$$

Now we inquire as to the pressure on the wall due to the impacts of all the $N/3$ molecules that strike it. If the average value of v^2 for the different molecules is designated by $\overline{v^2}$, the sum of the individual forces they exert is

$$F_{\text{total}} = \frac{N}{3}\frac{\overline{mv^2}}{L}. \tag{A–6}$$

The pressure on the wall is this force divided by the area L^2 of the wall:

$$\text{Pressure} = \frac{\text{total force on wall}}{\text{area of wall}},$$

$$p = \frac{F_{\text{total}}}{L^2} = \frac{1}{3}\frac{N\overline{mv^2}}{L^3}.$$

But L^3 is the volume V of the box! Hence

$$p = \frac{1}{3}\frac{N\overline{mv^2}}{V},$$

or, multiplying both sides by V, we have

$$pV = \frac{1}{3}N\overline{mv^2}. \tag{A-7}$$

A–4 MOLECULAR KINETIC ENERGY

The significance of Eq. (A–7) becomes clearer when we rewrite it in the form

$$pV = \tfrac{2}{3}N(\overline{\tfrac{1}{2}mv^2}) = \tfrac{2}{3}N\,\overline{\text{KE}}, \tag{A-8}$$

where $\overline{\text{KE}} = (\overline{\tfrac{1}{2}mv^2})$ is the average kinetic energy per gas molecule. This is the closest that the kinetic theory of gases can come to the ideal gas law,

$$pV = nRT. \tag{A-4}$$

The two formulas agree, however, if we assume that

The average kinetic energy of the molecules of a gas is proportional to the absolute temperature of the gas.

This assumption means that we set Eqs. (A–8) and (A–4) equal to each other. Then we have

$$\tfrac{2}{3}N\,\overline{\text{KE}} = nRT$$

and

$$\overline{\text{KE}} = \frac{3}{2}\frac{R}{(N/n)}T.$$

Now N is the number of molecules and n is the number of kmoles in the sample, so their ratio, N/n, is equal to the number of molecules in a kmole, which is Avogadro's number N_0. The quantity

$$k = \frac{R}{N/n} = \frac{R}{N_0} = \frac{8.32 \times 10^3 \text{ j/kmole·}^\circ\text{K}}{6.025 \times 10^{26} \text{ molecules/kmole}}$$

$$= 1.38 \times 10^{-23} \text{ j/molecule·}^\circ\text{K}$$

is therefore the universal gas constant expressed in terms of molecules rather than in terms of kmoles; it is known as *Boltzmann's constant*, after the 19th-century Austrian physicist Ludwig Boltzmann. The symbol for Boltzmann's constant is k, and its value is

$$k = 1.38 \times 10^{-23} \text{ j/}^\circ\text{K}.$$

In terms of Boltzmann's constant,

$$\overline{KE} = \tfrac{3}{2}kT; \tag{A-9}$$

$\tfrac{3}{2}kT$ *is the average kinetic energy of the molecules of all gases at the absolute temperature* T.

The interpretation of absolute zero in terms of the elementary kinetic theory of gases is a simple one: it is that temperature at which all molecular movement ceases.

The kinetic theory of gases leads us directly to the ideal gas law, but the ideal gas law is only a good approximation to reality. If we examine the initial assumptions we made in order to obtain the kinetic theory, it is easy to see why we should expect discrepancies between theory and experiment. For example, we assumed tacitly that gas molecules have volumes so small as to be negligible, that they exert no forces upon one another except in actual collisions, that these collisions conserve kinetic energy, and that the molecules have no kinetic energy of rotation or internal vibration. When the kinetic theory is worked out starting from more realistic assumptions, the results are in excellent agreement with observational data.

A-5 MOLECULAR SPEEDS

We can use Eq. (A-9) to compute the average speed of gas molecules whose mass is m:

$$\tfrac{1}{2}mv^2 = \tfrac{3}{2}kT,$$

$$v_{rms} = \sqrt{\overline{v^2}} = \sqrt{\frac{3kT}{m}}. \tag{A-10}$$

The above speed is denoted v_{rms} because it is the square root of the mean of the squared molecular speeds—the "root-mean-square" speed—and therefore different from the simple arithmetical average speed $\bar{v}$. To emphasize their difference with a simple example, we can evaluate both kinds of average for an assembly of two molecules, one with a speed of 1 m/sec and the other with a speed of 3 m/sec. We find that

$$\bar{v} = \frac{v_1 + v_2}{2} = \frac{(1+3)\text{ m/sec}}{2} = 2 \text{ m/sec},$$

whereas

$$v_{rms} = \sqrt{\frac{v_1^2 + v_2^2}{2}} = \sqrt{\frac{1^2 + 3^2}{2}} \text{ m/sec} = \sqrt{5} \text{ m/sec} = 2.24 \text{ m/sec},$$

so that v_{rms} and $\bar{v}$ are not the same. The relationship between $\bar{v}$ and v_{rms} depends upon the specific variation in molecular speeds under con-

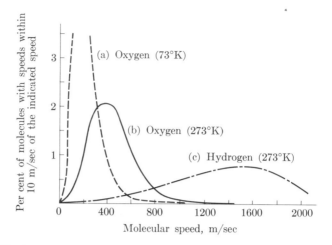

FIG. A–5. The distribution of molecular speeds in oxygen at 73°K (−200°C), in oxygen at 273°K (0°C), and in hydrogen at 273°K. The average molecular speed increases with temperature and decreases with molecular weight; hence molecular speeds in oxygen at 73°K (curve a) are on the average less than at 273°K (curve b), and at the same temperature molecular speeds in hydrogen (curve c) are on the average greater than in oxygen (curve b), since oxygen molecules are the heavier. At the same temperature the average molecular energy is the same for all gases.

sideration. For the distribution of molecular speeds found in a gas,

$$v_{\mathrm{rms}} \approx 1.09\,\bar{v},$$

so that the root-mean-square speed of Eq. (A–10) is about 9 percent greater than the arithmetical average speed $\bar{v}$.

Let us apply Eq. A–10 to the molecules of oxygen, one of the major constituents of air. Oxygen molecules are composed of two oxygen atoms each. The molecular weight of oxygen is therefore 2×16 or 32. As we learned earlier, one atomic weight unit is equivalent to a mass of 1.660×10^{-27} kg, and so each oxygen molecule has a mass of

$$m_{\mathrm{oxygen}} = 32 \times 1.660 \times 10^{-27}\,\mathrm{kg}$$

$$= 5.31 \times 10^{-26}\,\mathrm{kg}.$$

At an absolute temperature of 273°K (corresponding to 0°C), the root-mean-square speed of oxygen molecules is therefore

$$v_{\mathrm{rms}} = \sqrt{\frac{3 \times 1.38 \times 10^{-23}\,\mathrm{j/°K} \times 273°K}{5.31 \times 10^{-26}\,\mathrm{kg}}}$$

$$= 4.61 \times 10^2\,\frac{\mathrm{m}}{\mathrm{sec}},$$

which is a little over 1000 mi/hr! Evidently molecular speeds are very large compared with those of the macroscopic bodies familiar to us.

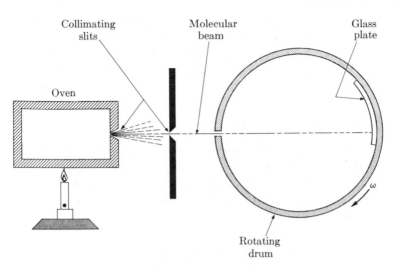

FIG. A-6. Rotating-drum method for determining molecular velocity distribution.

It is important to keep in mind that actual molecular speeds vary considerably on either side of v_{rms}. Figure A–5 is a graph showing the actual distribution of molecular speeds in oxygen at 73°K and at 273°K, and in hydrogen at 273°K. The basic shape of these curves is the same, and is given by the formula

$$N_v \, dv = 4\pi N \left(\frac{m}{2\pi kT}\right)^{3/2} v^2 e^{-mv^2/2kT} \, dv. \qquad (A–11)$$

Here $N_v \, dv$ is the number of molecules whose speeds are between v and $v + dv$, N is the total number of molecules present, m is the molecular mass, T is the absolute temperature, and k is Boltzmann's constant. Equation (A–11) was obtained a century ago by James Clerk Maxwell on the assumption of wholly random molecular motions with frequent collisions. The curves of Fig. A–5 are not symmetrical because the lower limit to v is fixed at $v = 0$ while there is, in principle, no upper limit; actually, as the curves show, the likelihood of speeds many times greater than v_{rms} is small.

The Maxwell distribution of molecular speeds can be verified directly in several ways. One ingenious method is illustrated in Fig. A–6. An oven is used to heat a sample of bismuth to 800°C. At this temperature bismuth is a gas, and it escapes through a slit in the oven and is collimated by a second slit some distance away. Near the second slit is a drum that spins at 6000 rev/min. When a slit in the drum faces the bismuth beam, a group of molecules enter the drum together. These molecules have different speeds, and accordingly arrive at the opposite side of the drum at different times. Owing to the rotation of the drum the fast and slow molecules are deposited at different locations. By lining the drum with a glass plate the distribution of deposited bismuth can be measured and from it can be determined the distribution of molecular speeds in the bismuth beam. The results agree with the predictions of Eq. (A–11).

A–6 KINETIC THEORY OF MATTER

The kinetic theory of gases does not meet with very much quantitative success when its notions are applied to the solid and liquid states. However, the concept that the heat energy of a body resides in the kinetic energy of its molecules helps in understanding a variety of phenomena in these states.

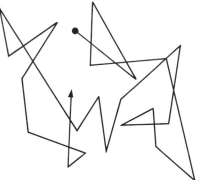

FIG. A–7.   Brownian motion.

The random motion of water molecules led to an important event in the history of science. In 1827 the British botanist Robert Brown noticed that pollen grains in water are in continual, agitated movement. Similar *Brownian motion* is apparent whenever very small particles are suspended in a fluid medium, for example smoke particles in air (Fig. A–7). According to kinetic theory, Brownian motion originates in the bombardment of the particles by molecules of the fluid. This bombardment is completely random, with successive molecular impacts coming from different directions and contributing different impulses to the particles. Albert Einstein, in 1905, found that he could account for Brownian motion quantitatively by assuming that, as a result of continual collisions with fluid molecules, the particles themselves have an average kinetic energy of $\frac{3}{2}kT$. Surprising as it may seem, this was the first direct verification of the reality of molecules, and it convinced many distinguished scientists who had previously been reluctant to believe that such things actually exist.

Another kinetic-molecular phenomenon characteristic of the liquid state is evaporation. A dish of water well below its boiling point of

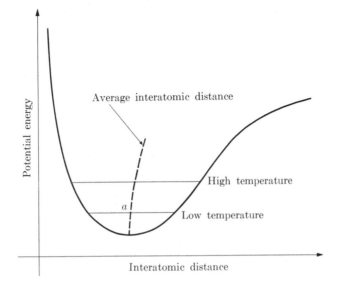

FIG. A–8. The atomic potential energy of a crystal as a function of the spacing of its constituent atoms. The increase in the average interatomic distance with energy is the cause of thermal expansion in a solid.

100°C will nevertheless gradually turn into vapor, growing colder as it does so. The faster the evaporation, the more pronounced the cooling effect; alcohol and ether chill the skin upon contact because of their extreme volatility. This behavior follows from the distribution of molecular speeds in a liquid. While not identical with that found in a gas, this distribution resembles those shown in Fig. A–5 in that a certain fraction of the molecules in any sample have much greater and much smaller velocities than the average. The fastest molecules have enough energy to escape through the liquid surface despite the attractive forces of the other molecules. The molecules left behind redistribute the available energy in collisions among themselves, but, because the most energetic ones escape, the average energy that remains is less than before and the liquid is now at a lower temperature.

Thermal expansion in a solid also has a straightforward explanation in terms of kinetic theory. Most solids are crystalline in nature, which means that the various atoms that compose them form a regular arrangement in space. (In some crystalline solids the basic constituents are whole molecules, rather than individual atoms, but we shall refer to them as atoms in general here for convenience.) The atoms behave as though they are joined together by tiny springs, thereby accounting for Hooke's law, and constantly oscillate about their equilibrium positions. Figure A–8 shows how the atomic potential energy of a solid varies with interatomic spacing. The normal spacing a corresponds to the lower portion of the curve, where the energy per atom is least. The amplitude of the vibrations is determined by the width of the curve: when the atomic separation is a minimum or a maximum, the energy of a pair of adjacent atoms is wholly potential, as in the case of a harmonic oscillator at each end of its path, while in the middle their energy is wholly kinetic. The average interatomic spacing is what determines a and hence the dimensions of the solid. When additional energy of thermal origin is added, the atomic spacing alternates through a wider range than before. If the potential energy curve were symmetrical about the distance a, no change in the dimensions of the solid would occur, since a would always be halfway between the two parts of the curve. However, the attractive and repulsive forces between atoms vary with distance in different ways, with the repulsive force increasing more rapidly as the atoms move closer together than the attractive force increases as the atoms move farther apart. Consequently, as shown in Fig. A–8, the average interatomic spacing becomes larger as the thermal energy of the solid increases and the amplitudes of the atomic vibrations increase. As it happens, changes in the average interatomic spacing are very nearly proportional to changes in temperature, leading to a linear thermal expansion formula.

Generally speaking, the stronger the atomic binding forces, the steeper and narrower the potential energy curve is, and hence the less pronounced the thermal expansion. As we have seen in Chapter 7, covalent and ionic bonds are very strong, metallic bonds intermediate, and molecular bonds (in which whole molecules constitute the crystal) weakest. Accordingly we are not surprised to find that the coefficient of thermal expansion of diamond, a covalent solid, is $0.11 \times 10^{-5}/°C$, that of iron, a metal, is $1.2 \times 10^{-5}/°C$, and that of ice, a molecular solid, is $5.1 \times 10^{-5}/°C$. Some solids do not have their atoms arranged symmetrically, and their potential energy curves are different in different directions. Thus the expansion coefficient of a quartz crystal parallel to a certain direction is $1.4 \times 10^{-5}/°C$, while it is $0.9 \times 10^{-5}/°C$ perpendicular to this direction. ("Fused quartz" is a glasslike noncrystalline form of quartz which has the very low thermal expansion coefficient of $0.05 \times 10^{-5}/°C$ in all directions.)

Appendix B
DUMBBELL ROTATOR

An object that consists of two particles at either end of a weightless rod—a sort of idealized dumbbell—is a reasonably good approximation of both a one-electron atom and a diatomic molecule. Such an object can rotate about its center of mass, and we require a formula for its moment of inertia when we consider the dynamics of atoms and molecules. The center of mass of the system is determined by the condition that

$$m_1 r_1 = m_2 r_2, \tag{B-1}$$

where r_1 and r_2 are respectively the distances of the particles of masses m_1 and m_2 from the center of mass (Fig. B–1). If the particles are a distance l apart,

$$r_1 + r_2 = l. \tag{B-2}$$

The moment of inertia of the system about an axis through the system's center of mass perpendicular to the line joining the particles is

$$I = m_1 r_1^2 + m_2 r_2^2. \tag{B-3}$$

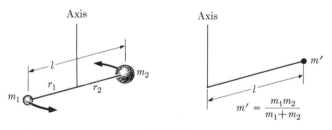

FIG. B–1

We can obtain a more convenient expression for I by proceeding as follows. First we multiply Eq. (B–3) by $(m_1 + m_2)/(m_1 + m_2)$ to obtain

$$I = \frac{m_1^2 r_1^2 + m_1 m_2 r_1^2 + m_1 m_2 r_2^2 + m_2^2 r_2^2}{m_1 + m_2}. \tag{B-4}$$

From Eq. (B–1) we see that $m_1 = m_2 r_2/r_1$ and $m_2 = m_1 r_1/r_2$. We can therefore substitute

$$m_1^2 = m_1 m_2 \frac{r_2}{r_1} \quad \text{and} \quad m_2^2 = m_1 m_2 \frac{r_1}{r_2}$$

in the first and last terms of the numerator of Eq. (B–4), with the result that

$$I = \frac{m_1 m_2}{m_1 + m_2} (r_1^2 + 2r_1 r_2 + r_2^2) = \frac{m_1 m_2}{m_1 + m_2} (r_1 + r_2)^2$$

$$= \frac{m_1 m_2}{m_1 + m_2} l^2 = m' l^2. \tag{B–5}$$

The quantity

$$m' = \frac{m_1 m_2}{m_1 + m_2} \tag{B–6}$$

is known as the *reduced mass* of the system of two particles because m' is smaller than either m_1 or m_2. According to Eq. (B–5), a pair of rigidly connected particles that are the distance a apart have a moment of inertia about their center of mass equal to that of a single particle of mass m' that rotates the distance a from its axis of rotation.

Appendix C

TWO-BODY OSCILLATOR

Figure C–1 shows two bodies of masses m_1 and m_2 that are joined by a spring. The bodies are on a horizontal, frictionless surface so that the only forces that affect their motions are those the spring exerts. In the absence of external forces the linear momentum of the system must remain constant, and the oscillations of the bodies therefore cannot affect the motion of their center of mass. Thus m_1 and m_2 vibrate back and forth relative to their center of mass in opposite directions: if m_1 has the velocity $\mathbf{v}_1$ at some instant, m_2 must have a velocity $\mathbf{v}_2$ at that instant such that

$$m_1\mathbf{v}_1 + m_2\mathbf{v}_2 = 0,$$

$$\mathbf{v}_2 = -\frac{m_1}{m_2}\mathbf{v}_1. \tag{C-1}$$

When m_1 moves to the left, m_2 moves to the right, and both reach the respective extremes of their motions at the same time. What is the frequency of these oscillations?

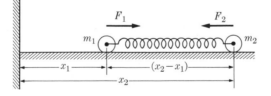

FIG. C–1.   A two-body oscillator.

It is convenient to examine the problem using the coordinate system of Fig. C–1. The length of the spring at any time is $x_2 - x_1$, which equals the normal length l of the spring plus its displacement x:

$$l + x = x_2 - x_1,$$

$$x = x_2 - x_1 - l. \tag{C-2}$$

The displacement x is positive when the spring is extended and negative when it is compressed. The forces F_1 and F_2 which the spring exerts on each body are equal in magnitude and opposite in direction, and so

$$F_1 = kx, \qquad F_2 = -kx, \tag{C-3}$$

where k is the force constant of the spring. From the second law of motion, $\mathbf{F} = m\mathbf{a}$, we find that the bodies must obey the equations

$$m_1 a_1 = m_1 \frac{d^2 x_1}{dt^2} = kx, \tag{C-4}$$

$$m_2 a_2 = m_1 \frac{d^2 x_2}{dt^2} = kx. \tag{C-5}$$

These equations can be combined by multiplying the first by m_2 and the second by m_1 and then subtracting the first from the second. This procedure yields

$$m_1 m_2 \frac{d^2 x_1}{dt^2} - m_1 m_2 \frac{d^2 x_2}{dt^2} = -m_1 kx - m_2 kx,$$

which can be rewritten

$$\frac{m_1 m_2}{m_1 + m_2} \left(\frac{d^2 x_1}{dt^2} - \frac{d^2 x_2}{dt^2} \right) = -kx. \tag{C-6}$$

Now we differentiate Eq. (C-2) twice with respect to time. Since l, the normal spring length, is a constant, we obtain

$$\frac{d^2 x}{dt^2} = \frac{d^2 x_1}{dt^2} - \frac{d^2 x_2}{dt^2}.$$

Equation (C-6) can therefore be stated entirely in terms of x, the spring displacement:

$$\frac{m_1 m_2}{m_1 + m_2} \frac{d^2 x}{dt^2} = -kx,$$

$$m'a = -kx. \tag{C-7}$$

The quantity

$$m' = \frac{m_1 m_2}{m_1 + m_2} \tag{C-8}$$

is called the *reduced mass* of the system, and a is the acceleration a particle of mass m' would have if it were attached to a spring of force constant k which is displaced by x from its equilibrium position. Hence the motion of a two-body oscillator is described by the same equation (Eq. C-7) as that of a single body of mass m' attached to the same spring.

We therefore conclude that the frequency of the two-body oscillator is

$$f = \frac{1}{2\pi}\sqrt{\frac{k}{m'}}. \tag{C-9}$$

The notion of the reduced mass of a system of two particles was also introduced in Appendix B, where we saw that the rotation of two linked particles separated by a distance l about their center of mass is equivalent to that of a single particle of mass m' rotating about an axis at a distance l from it.

The relative energies of the two vibrating bodies are not difficult to determine. If v_{1m} and v_{2m} are the maximum speeds of m_1 and m_2, the ratio of their total energies (kinetic and potential, since PE = 0 when the speeds are maxima) is

$$\frac{E_1}{E_2} = \frac{\frac{1}{2}m_1 v^2{}_{1m}}{\frac{1}{2}m_2 v^2{}_{2m}}.$$

From Eq. (C-1), $v_{2m} = -(m_1/m_2)v_{1m}$, and

$$\frac{E_1}{E_2} = \frac{m_2}{m_1}. \tag{C-10}$$

The energies of the oscillating bodies are inversely proportional to their masses: the heavy body has *less* energy than the lighter one.

Appendix D

THE ELEMENTS

Atomic number	Element	Symbol	Atomic weight*	Atomic number	Element	Symbol	Atomic weight*
1	Hydrogen	H	1.008	52	Tellurium	Te	127.6
2	Helium	He	4.003	53	Iodine	I	126.9
3	Lithium	Li	6.939	54	Xenon	Xe	131.3
4	Beryllium	Be	9.012	55	Cesium	Cs	132.9
5	Boron	B	10.81	56	Barium	Ba	137.3
6	Carbon	C	12.01	57	Lanthanum	La	138.9
7	Nitrogen	N	14.01	58	Cerium	Ce	140.1
8	Oxygen	O	16.00	59	Praseodymium	Pr	140.9
9	Fluorine	F	19.00	60	Neodymium	Nd	144.2
10	Neon	Ne	20.18	61	Promethium	Pm	(147)
11	Sodium	Na	22.99	62	Samarium	Sm	150.4
12	Magnesium	Mg	24.31	63	Europium	Eu	152.0
13	Aluminum	Al	26.98	64	Gadolinium	Gd	157.3
14	Silicon	Si	28.09	65	Terbium	Tb	158.9
15	Phosphorus	P	30.97	66	Dysprosium	Dy	162.5
16	Sulfur	S	32.06	67	Holmium	Ho	164.9
17	Chlorine	Cl	35.45	68	Erbium	Er	167.3
18	Argon	A	39.95	69	Thulium	Tm	168.9
19	Potassium	K	39.10	70	Ytterbium	Yb	173.0
20	Calcium	Ca	40.08	71	Lutetium	Lu	175.0
21	Scandium	Sc	44.96	72	Hafnium	Hf	178.5
22	Titanium	Ti	47.90	73	Tantalum	Ta	180.9
23	Vanadium	V	50.94	74	Tungsten	W	183.9

Number	Name	Symbol	Atomic Weight
24	Chromium	Cr	52.00
25	Manganese	Mn	54.94
26	Iron	Fe	55.85
27	Cobalt	Co	58.93
28	Nickel	Ni	58.71
29	Copper	Cu	63.54
30	Zinc	Zn	65.37
31	Gallium	Ga	69.72
32	Germanium	Ge	72.59
33	Arsenic	As	74.92
34	Selenium	Se	78.96
35	Bromine	Br	79.91
36	Krypton	Kr	83.80
37	Rubidium	Rb	85.47
38	Strontium	Sr	87.62
39	Yttrium	Y	88.91
40	Zirconium	Zr	91.22
41	Niobium	Nb	92.91
42	Molybdenum	Mo	95.94
43	Technetium	Tc	(99)
44	Ruthenium	Ru	101.1
45	Rhodium	Rh	102.9
46	Palladium	Pd	106.4
47	Silver	Ag	107.9
48	Cadmium	Cd	112.4
49	Indium	In	114.8
50	Tin	Sn	118.7
51	Antimony	Sb	121.8
75	Rhenium	Re	186.2
76	Osmium	Os	190.2
77	Iridium	Ir	192.2
78	Platinum	Pt	195.1
79	Gold	Au	197.0
80	Mercury	Hg	200.6
81	Thallium	Tl	204.4
82	Lead	Pb	207.2
83	Bismuth	Bi	209.0
84	Polonium	Po	210
85	Astatine	At	(210)
86	Radon	Rn	222
87	Francium	Fr	(223)
88	Radium	Ra	226.1
89	Actinium	Ac	227
90	Thorium	Th	232.0
91	Protactinium	Pa	231
92	Uranium	U	238.0
93	Neptunium	Np	(237)
94	Plutonium	Pu	(242)
95	Americium	Am	(243)
96	Curium	Cm	(245)
97	Berkelium	Bk	(249)
98	Californium	Cf	(249)
99	Einsteinium	Es	(253)
100	Fermium	Fm	(255)
101	Mendelevium	Md	(256)
102	Nobelium	No	(253)
103	Lawrencium	Lw	(257)

*Those in parentheses refer to the most stable isotope of the element.

Appendix E

SELECTED PHYSICAL CONSTANTS

Quantity	Symbol	Value
Absolute zero	$0°K$	$-273°C$
Avogadro's number	N_0	6.025×10^{26} kmoles^{-1}
Boltzmann's constant	k	1.38×10^{-23} j/$°K$
Electrostatic constant*	ϵ	8.85×10^{-12} coul/n·m^2
Electromagnetic constant	μ	1.26×10^{-6} weber/amp·m
Charge of electron	e	1.60×10^{-19} coul
Electron rest mass	m_e	9.11×10^{-31} kg
Proton rest mass	m_p	1.67×10^{-27} kg
Neutron rest mass	m_n	1.67×10^{-27} kg
Speed of light	c	3.00×10^8 m/sec
Planck's constant	h	6.63×10^{-34} j·sec

$$* \quad \frac{1}{4\pi\epsilon_0} = 9 \times 10^9 \text{ n·m}^2/\text{coul}^2$$

Answers to Odd-Numbered Problems

CHAPTER 1

1. A vertical antenna will give better reception because the electric vector of the radiation is vertical; therefore the induced emf in a vertical antenna will be greater.
3. 10^{-11} w/m^2 5. 2.4×10^{-7} n
7. 3.25×10^5 km
9. 3.3×10^{-13} w/m^2; 1.3×10^{-11} watts/m^2; 4.4×10^{-20} n/m^2
11. Because momentum is a vector quantity and the sun radiates equally in all directions.
13. 4.4×10^8 n/m^2; 4.4×10^3 atm

CHAPTER 2

1. No; yes; yes   5. 2.6×10^8 m/sec   9. 3.7×10^{-12} kg
3. 1:40.3; 1:40.3   7. 4.2×10^7 m/sec   11. 6.05×10^{-8} %
13. 2.6×10^8 m/sec   15. 4.2×10^7 m/sec; 9.3×10^6 ev
17. Nothing material moves across the screen.

CHAPTER 3

1. Even a faint light involves a great many photons.
3. 1.71×10^{31} photons/sec   5. 4.64×10^4 sec^{-1}
7. Diffraction, interference, and polarization can be exhibited much more simply than the photoelectric or other quantum effect.
9. 4.14×10^4 v   11. 2.41×10^{18} sec^{-1}; x-rays
13. 9.95×10^{15} j; 3.32×10^{-23} kg·m/sec
15. 9.8×10^{18} sec^{-1}; 1.5×10^{-16} j
17. 1.34×10^{19} sec^{-1}; 1.10×10^{19} sec^{-1}
19. 2.73×10^{-39} m   21. 2.87×10^{-14} m
23. Apply electric or magnetic field and look for deflection, or try Compton-type experiment.
25. 7.3×10^5 m/sec; 4.0×10^2 m/sec

CHAPTER 4

1. They agree that the positive matter in the atom has much greater mass than the negative matter, but in the Thomson model the positive matter occupies the entire atomic volume, while it occupies only a tiny region at the center of the atom in the Rutherford model.

3. 2.2×10^6 m/sec
5. A hydrogen sample contains a great many atoms, each of which has a variety of possible transitions.
7. 12 ev
9. 4.05×10^{-6} m   11. 8.3×10^{-20} j; 0.52 ev   13. Lyman series
15. 9.73×10^{-8} m; 6.81×10^{-27} kg·m/sec; 4.08 m/sec
17. Emission

CHAPTER 5

3. 2
7. The outer shells of these atoms all have either 1 or 2 electrons, since the increases in atomic number in this sequence go into filling up the inner $n-3$, $l-2$ subshell.
9. Both Li and Na atoms have one electron outside a closed shell.
11. Lithium atoms must *lose* an electron each in forming a molecule, and sharing electrons does not permit this to occur.

CHAPTER 6

1. 0.5 ev; 2.9×10^{-9} m
3. Li atoms have one electron outside a closed shell, which is more easily removed than any of the electrons of F atoms whose outer shells lack an electron of completion.
5. Na^+ ions have closed shells, while Na atoms have a single, easily detached outer electron.
7. All rare gas atoms contain only closed electron shells.
9. 26.8 amp·hr   11. 1.73 g
13. 1.03 kg   15. 27.6 hr
17. 2.23×10^{-10} m   19. 8   21. 8.85×10^{11} sec^{-1}

CHAPTER 7

1. Ionic, NaCl; covalent, diamond; van der Waals, ice; metallic, iron. The interaction in each case is electrostatic.
3. Van der Waals
5. "Holes," which are sites of missing electrons
7. Both are nonconductors at very low temperatures; germanium is the better conductor at room temperature, because its forbidden band is narrower.
9. 1.5 R
11. 4.26×10^4 °K; 1.39×10^6 m/sec

CHAPTER 8

1. 2500 v   3. 6.1 mm; 81% B^{11}, 19% B^{10}
5. 5 p, 5 n; 10 p, 12 n; 16 p, 20 n; 38 p, 50 n; 72 p, 108 n
7. 289 Mev   9. 18.8 Mev; 19.4 Mev
11. 0.85 Mev; difference in binding energy is 0.62 Mev, so nuclear forces cannot be strongly charge-dependent.